Life before Man

Zdeněk V. Špinar

Life
before Man

illustrated by Zdeněk Burian

THAMES AND HUDSON · LONDON

Z. V. Špinar
Life before Man
Illustrated by Z. Burian under the direction of J. Augusta,
Z. V. Špinar, and V. Mazák

Special consultant: Dr Ian Cornwall,
Institute of Archaeology, London

Introductory text and captions by Z. V. Špinar

Translated by Margot Schierlová
Line drawings: A. Benešová

First published in Great Britain in 1972 by
Thames and Hudson Ltd, London

Designed and produced by Artia for Thames and Hudson

Printed in Czechoslovakia by Polygrafia, Prague

ISBN 0 500 01089 7

Contents

Preface

The aim of this book is to introduce the reader to the incredible and fascinating panorama of life on earth as it has unfolded from its earliest beginnings more than 3000 million years ago to the arrival of *Homo sapiens* and the introduction of settled farming, a mere 5000 years ago. The main text, the first section of the book, sets out the key events in that story; the colour illustrations that follow attempt to recreate as accurately and as vividly as possible the actual appearance of living things in their contemporary environments. In this way the reader may imaginatively visualize the whole past life of the world, a past that merges imperceptibly with the living present of which he himself is a part. I hope that a greater understanding of the earth's history will make the reader more deeply aware of his own world today, and so encourage him to feel a keener responsibility for the future development of life on our planet.

Zdeněk V. Špinar

Birth of the solar system

The beginnings of our planet and of the solar system as a whole are veiled by the impenetrable curtain of time. For centuries, the question of the origin of the earth was monopolized by philosophers, because nobody could provide a factual explanation for certain of its phenomena. It was not until the 18th century that the first scientific hypotheses on the origin of the earth and the solar system, based on astronomical observations, were proposed. Since then there have been many of these theories, the earlier ones being constantly modified or replaced as knowledge increases.

The first was the famous theory proposed by the German philosopher Immanuel Kant in 1775. Kant believed that the whole of the solar system was formed from particles of a primordial matter originally scattered throughout space. These particles moved in different directions, collided with each other, and so lost velocity. Heavier and more compact particles were attracted to each other by the force of gravity, and gradually combined to form a central body, the sun, which then drew toward itself further, lighter, and smaller particles. This gave rise to numbers of spinning masses crossing each others' course. Some of these masses, which originally often moved in opposite directions, eventually "fell into step" and began to circle the sun as rings, in roughly one plane, without interfering with each other in any way. Each separate ring of gaseous matter had a more solid core, and the lighter particles clustered round this in a spherical mass, thereby slowly giving rise to the planets, which orbit the sun in the same plane as the original rings.

Laplace's nebular theory

In 1796 the French mathematician and astronomer Pierre Simon Laplace proposed a somewhat different theory. Laplace suggested that the sun already existed as a vast, glowing, gaseous mist (nebula) of low density but huge size. He assumed that it spun freely in space, but that as the force of gravity caused it to contract, it rotated faster and faster until centrifugal forces flattened it to the shape of a slightly bulging disk. Because of the increase in the sun's rate of rotation, centrifugal force overcame gravity at its equatorial margin, so that matter there broke away from the disk to form a ring. As the nebula

7

went on shrinking and spinning, other rings were successively detached, and condensed to form the various planets. Altogether 10 rings broke away, forming the nine planets and the asteroids (a belt of minor planets). The satellites of the individual planets in turn developed from secondary rings formed from the matter of the planets while they still consisted of very hot gas.

As their matter condensed, the newly formed bodies grew intensely hot. Our earth also once consisted of incandescent gas and shone like a star. But gradually it cooled and liquefied, and later, in the next stage of cooling, acquired a solid crust. The crust was enveloped in a heavy atmosphere, from which further cooling extracted the earth's water.

These two theories complement each other, so they are often put together and described as the Kant-Laplace theory. Since later scientists had no better explanation to offer, this theory won many adherents during the 19th century.

Modern theories

Among other theories we find "catastrophe" hypotheses, which attribute the formation of the earth to some outside interference; for instance, to the passage of a star near the sun, causing some of its matter to be detached and scattered. As a result of expansion, the glowing gaseous material quickly cooled and then shrank, producing large numbers of solid particles that later came to form the rudiments of the planets.

In recent years, new ideas have been put forward by American and Russian scientists. Whereas the development of the earth was formerly regarded as a single, unbroken cooling process, the new theories presume that a number of opposing processes occurred. They suppose that, together with cooling (which causes loss of energy), processes that produce new energy and compensate for the gradual loss of heat are also taking place in the earth. The author of one of these modern hypotheses, the American astronomer F. L. Whipple, called his hypothesis (1948) the dust-cloud theory, but it is actually only an extended version of the Kant-Laplace nebular theory.

With modern improvements in technical equipment and greater knowledge of the chemical composition of the solar system, astronomers returned to the idea that the sun and planets were formed from a huge (but cold) nebula consisting of gas and dust. Powerful telescopes have detected many such enormous clouds of gas and

8

dust far out in space, some of them in the actual process of condensing to form new stars. The original Kant-Laplace theory is thus gradually re-elaborated, because, supported by new evidence, it can still be used to explain the origin of the solar system.

Every theory about the origin of the universe has made some new contribution to the complex problem of the origin of the earth. But all of them show that the formation of the earth and of the solar system is bound up with the development of the stars and of the universe as a whole. The earth was formed along with a group of other planets circling the sun, and together they constitute the most important parts of the solar system.

The earth: atmosphere and hydrosphere

The earth's shape has probably not changed at all since the planet was first formed about 4600 million years ago. Its chemical structure is also fundamentally the same, although the distribution of its 90-odd elements has altered considerably. Another difference is that the earth's surface was originally plain and unmarked by erosion.

The earth's original atmosphere was formed from interstellar gas and consisted chiefly of hydrogen and helium; but the force of gravity was not strong enough to retain light gases in the initial phases of the earth's development, so they escaped into space and were expelled from the solar system.

The earth's present atmosphere is of secondary origin. It was, and is still, being formed from gases produced by organic activity on the surface and by volcanic processes within the earth. For a relatively short time the earth was virtually without water, but it acquired a hydrosphere (the oceans, seas and other waters) in the same way as an atmosphere. And because the earth lies at a convenient distance from the sun (about 93 million miles), the temperature at its surface varied only within a fairly narrow range, generally slightly above 0°C, so that water remained liquid. This was very important for the further history of the earth, because water is one of the best and most convenient media for accommodating a wide range of chemical reactions. As soon as the earth's surface was able to retain water, which formed continuous sheets in any depressions, a new phase in the earth's development, known as the oceanic age, started.

Violent atmospheric storms and heavy rainfalls became frequent. These dissolved all soluble salts lying on the exposed surface, or extracted them by percolation through the rocks, and the solutions started to concentrate in the primeval oceans. Seawater was thus salt from the very outset. With the development of the atmosphere and hydrosphere, new factors shaping the face of the earth appeared.

Sedimentary rocks

The earliest known rocks are crystalline and have solidified on cooling from molten rock material called magma. In those parts of the earth not covered by water such primeval surface rocks were subjected to physical and chemical weathering, and so began to crack and crumble. Water and wind carried the particles away and deposited them in different places in the form of sediments.

These sediments were successively deposited in layers, often at the bottom of an ocean. Folding of the earth's surface occurred several times. New continents were upheaved here and sank again there. New rocks were formed and then destroyed again. Sediments furnish evidence of this development. We can tell if they were deposited in a sea or a desert, in a warm, dry period or in an ice age; and we can even determine just when they were formed. If we could find a place where the sedimentary layers had been left intact, just as they were deposited through all the geological periods down to the present day, they would give us a complete picture of the earth's history. That picture would include the history of life, because the various layers contain the fossilized remains of animals and plants living at the time when they were laid down. These tell us how life on the earth developed and what the ancestors of present-day animals and modern man looked like.

We know from geology that the earth's crust is never still. Some parts are being lifted, others are sinking. In some places the sea is receding, giving rise to new areas of dry land; in others it is encroaching on the land, so that whole regions are steadily disappearing beneath the waves. The turbulent development of the earth's surface made it impossible for the various layers to be deposited in an unbroken series, so the number and character of the layers vary in different regions and are often incomplete.

1 As a result of physical and chemical weathering rocks disintegrate and crumble. Water and wind carry the loose debris to basins, where they are deposited in the form of sediments.

We get a complete survey of the earth's development only by studying the sequence of the layers in different parts of its surface and combining the results. This is part of historical geology, the main branch of which is the science of the layers (strata) of the earth's crust. This is known as stratigraphy or stratigraphic geology and is based largely on the study of the fossils preserved in the various layers.

The degree of development of these extinct organisms tells us the relative age of the layers and their place in the stratigraphic scale. The science that deals with the study of the fossils is called paleontology.

On the basis of stratigraphic studies, geologists have divided the earth's history into four geological eras of varying lengths, which they term Precambrian, Paleozoic, Mesozoic, and Cenozoic. These eras are further subdivided into various periods, as shown in the several geological time charts of the illustrated section (pp. 49—220). We shall start our story with the Precambrian era, the age of the primeval oceans and the time when life first appeared on the earth.

Precambrian era

This geological era starts with the formation of the earth's crust and ends with the appearance of animals with a hard body-covering. It is usually divided into the Archaeozoic period (the earliest geological period), which lasted about 2000 million years, and the Proterozoic period, which lasted about 1030 million years. It was at some point during this tremendously long span of time—four fifths of all geological time—that life on our planet originated.

Origins of life

Life was able to appear as soon as conditions, especially the temperature, were suitable. One of the conditions for life is the presence of chemical substances called proteins, but the temperature at the earth's surface first had to be below 95°C, the highest temperature at which living matter can exist. Certain carbon compounds, mainly some proteins, essential for building living tissue, can persist even at this high temperature. But it is hard to say how long it took before the temperature of the earth's surface reached the necessary level.

Most scientists studying the origin of life on the earth believe that it started in shallow seawater as a result of ordinary physical and

11

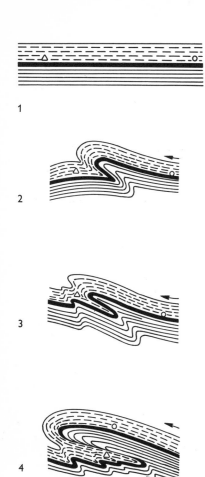

1

2

3

4

2 Pressures (black arrow) in the earth's crust deform horizontal sedimentary strata (1), causing them to yield by folding (2), overfolding and crumpling (3); finally, if the stress is maintained, the strata fracture and thrust (4). The relative movement of the triangular and circular symbols shows extent of shortening in crust.

chemical processes working on simple primary compounds. When the basic laws of chemistry were identified and defined, it was shown that chemical compounds are formed under given conditions and from given amounts (by weight) of the elements. It was found that the probability of forming the compounds of which living individuals are composed was high—because of the ordinary affinities of the chemical elements and their compounds—and that it was controlled by fundamental laws of chemistry and physics.

Molecules with the high degree of complexity needed to form living matter took a long time to develop. Organized living systems were synthesized from inert material. This is the first point at which we can speak of the development, or evolution, of the whole process that culminated in the formation of a living organism. The period of time needed to develop the elements, chemical compounds, and life was of immense duration. It goes back some 4600 million years, to the formation of the earth itself. The first step was the formation of the elements. The second, after the earth had been formed, was the combination of elements to make compounds. Finally, organic compounds, (substances made up mainly of carbon), complex multi-molecular formations, and eventually what can properly be described as living systems appeared on the earth's surface.

The first living organisms were, of course, very simple. But natural selection—the process whereby those forms of life that are best adapted to their environment survive, while the less efficient become extinct—gradually enabled more complex life-forms to emerge. The oldest primitive organisms, which developed, we suppose, about halfway through the Archaeozoic period, were not yet divided into animals and plants. These two basic groups emerged gradually, and their differentiation seems to have been completed about halfway through the Proterozoic period. They lived and died in the primeval oceans, and as their bodies often accumulated in particular areas, they already left distinct traces of their existence.

The first organisms could live only on organic matter. But when they had consumed all the organic substances in their environment, they were either doomed to die, or had to develop the ability to synthesize within themselves organic substances from inorganic sources, that is, from carbon dioxide and water. During development, some organisms (the plants) did, in fact, acquire the ability to absorb sunlight and use it to break down water into its elements. They used hydrogen to reduce carbon dioxide to sugars, and in turn employed these to synthesize other organic substances in their bodies. This resulted in the development of organisms capable of converting

12

inorganic to organic substances within themselves. We find their remains in the oldest sediments of the earth's crust. The complete process (known as photosynthesis) was one of the fundamental developments in the story of life on earth, for without it no further progress of any kind could have been made.

The first animals and plants were microscopic single-celled organisms. The association of single cells of the same type in colonies was a slight step forward, but it was not until multicellular organisms evolved that real and significant biological progress became possible. These organisms were composed of cells, or groups of cells, of different shapes and with different functions, and their presence led to the explosive evolution of living organisms and in turn to the development of more complex and diverse forms. Animal and vegetable life developed considerably during the early Proterozoic period. More advanced forms of algae (simple water-dwelling plants) appeared in the sea, where they sheltered the first hollow-bodied animals—mollusks and worms. The further development of life can be studied fairly easily from the remains of the hard parts of different types of organisms left in the layers of the earth's crust. These remains, which by chance and a fortunate chemical environment in the surrounding sediment have been preserved down to the present day, are known as fossils.

The oldest fossils

The oldest fossils—and hence the first evidence of life on earth—are bacteria-like organisms found in Precambrian sediments in South Africa. Their age is estimated at 3200 million years. They consist of organisms that are so small that they can be detected only by means of the electron microscope. Composed of organic matter, they are quite well preserved, and look very much like present-day bacilli. Their biologically organic nature was demonstrated by chemical tests. Further examples of ancient life were found in Minnesota (2700 million years), Rhodesia (2700 million years), on the Canadian-United States border (2000 million years), in northern Michigan (1000 million years), and elsewhere.

No remains of animals with a hard body-covering have been found in Precambrian layers, but we have evidence that a number of groups of different animals did exist at that time. The most primitive ancient animals had no calcified skeleton or any other body-support capable of fossilization, but imprints and sometimes fossil finds of large multicellular animals preserved in Precambrian layers prove

13

that such organisms must have existed. Finds are extremely rare, however. For instance, curious tuber-like formations found in Canadian limestones and given the name *Atikokania* were considered by scientists to be related to marine sponges. The existence of larger organisms, probably worms, is borne out by creeping and burrowing marks preserved on the buried surfaces of former seabeds. Paleontologists did not find the actual animals, but from these traces they were able to determine the type of animal that made them.

The remains of very peculiar animals were found in 1947 by the Australian scientist R.C. Spriggs in the Edicara Hills, about 280 miles north of Adelaide, South Australia. They were examined by another Australian, Dr Martin F. Glaessner, who found that most of them were completely unknown types of animals, without either shell or skeleton. Most of them came from the jellyfish group, while others belong to the segmented worms and related animals. Some of them had no parallel among either living or extinct species of animals, however, although all the finds represent animals that lived in water.

Until recently it was thought that the Edicara finds were Precambrian fossils, but modern dating methods showed that the layers in which they occurred were only about 550 million years old. This means that they belong to the early Cambrian period. Similar fossils, probably of the same age, have been found in South Africa.

Simple microscopic plants were much more abundant than animals in the Precambrian era, and the traces of their existence constitute the oldest known evidence of life on the earth. The oldest such material, which was the outcome of the activity of bacteria, primitive algae, and lower plants, comes from rocks about 2500 million years old. These consist of a type of limestone that is not itself derived from the skeletons of plants, but is made up of material precipitated near the plants from seawater during biochemical processes associated with the intake of organic nutrients.

Life thus originated in the Precambrian seas and divided into two main branches—animal and plant. The first primitive, simple forms developed into multicellular, relatively complex bodies and forms, some of which represent the origins of plants and animals that in later geological periods spread over the whole globe. Newly developed life forms crowded the shallow parts of the sea and penetrated into fresh waters, and many of them had already started to prepare for the next important step in evolution—colonization of the dry land.

Paleozoic era

No one cannot fully grasp just how long 345 million years actually is, but that is the length of time for which the next era in the history of earth, the Paleozoic era (also known as the Primary era) lasted. Geologists divided it into six periods, the oldest of which is the Cambrian, followed in turn by the Ordovician, Silurian, Devonian, Carboniferous, and Permian. The Paleozoic ("ancient life") started with tremendous oceanic floods, which followed the lifting up of vast continents at the end of the Precambrian era. Geologists believe that at that time there existed already a single huge continental block, known as Pangaea ("all earth"), and that it was completely surrounded by ocean. In time, this original continent broke up into separate parts, which formed the later, and, finally, the present-day continents. The several main fragments drifted apart until they finally took up their present positions. The first coherent hypothesis of continental drift was proposed by Alfred Wegener, a German geologist, in 1912. In his view, Pangaea first divided into two supercontinents—Laurasia in the northern and Gondwanaland in the southern hemisphere (see diagram, right). The space between them was occupied by a sea called Tethys. In the next phase, during the Silurian period, a large northern continent was formed as a result of crustal folding and upheaval—the Caledonian (Scotland) and Hercynian (Harz mountains) phases of mountain building. During the Devonian, the irregular surface of this landscape was overlaid by the products of disintegration under weathering of large mountain masses, the particles of which in the hot, dry climate of the time were coated red with iron oxides, as are some modern desert sands. This Devonian continent is therefore known as the Old Red Continent. Many new types of land plants developed on it; the remains of the first land-dwelling vertebrates have actually been found in one part of it. The Old Red Continent is thus of special importance for us. In the later part of the Paleozoic era, Gondwanaland, which included most of South America, almost the whole of Africa, Madagascar, India, and Antarctica, still persisted as a single land mass in the southern hemisphere.

At the end of the Paleozoic the sea receded and Hercynian mountain-building slowly stopped. The first primitive types of plants and animals also died out.

A gradual succession of different groups of plants took place during the Paleozoic era. Marine algae predominated during the first part,

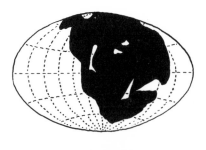
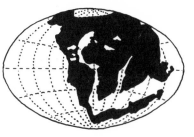

3 Alfred Wegener's theory of continental drift. Some 200 million years ago the continents were joined to form a single land mass. As this block split up, the fragments slowly drifted apart to occupy their present positions. The middle section shows the situation prevailing about 50 million years ago.

15

from the Cambrian to the Silurian period; later, up to the end of the Carboniferous period, spore-bearing plants were in the ascendancy. (Spores consist of millions of tiny detachable single cells that are distributed by wind and water; once lodged in a suitable habitat, they germinate and produce an intermediate, sexual, generation of plants, which in turn gives rise to spore-bearing, sexless, individuals.) At its end, especially in the second half of the Permian period, most of the plants were gymnosperms ("naked seeds") — plants whose seeds are not protected by seed vessels.

Plants conquer dry land

Apart from a few problematical exceptions, we know of no dry-land vegetation at the beginning of the Paleozoic era. But it is quite likely that some plants slowly started to penetrate inland at the end of the Precambrian, as many of the sediments of that time contained large and easily available supplies of nutrients. Before some types could become adapted to life on the dry land, however, they had to undergo fundamental and far-reaching anatomical changes. For example, they had to acquire a thick skin to prevent too rapid loss of moisture, and needed to develop a strong woody framework to help overcome the force of gravity, now that their weight was no longer borne up by water. They remained rooted in the ground, from which they drew nourishment and water, so that they needed tubular vessels to carry these substances to the parts of their bodies above ground. In addition, they needed fertile soil, the formation of which required the presence of numerous lower plant micro-organisms—bacteria, algae, fungi, and animal life. The products of the activity of these last, and their own dead bodies, gradually converted weathered crystalline and sedimentary rocks to rich soil capable of providing nourishment for more advanced plants. Attempts to colonize the dry land were increasingly successful, until perennial types of terrestrial plants developed. Evidence of this important advance has been preserved in Silurian marine sediments in central Bohemia, where the first well-preserved remains of more advanced plants of the Psilophytales ("naked plants," that is, without leaves) group can be found. Except for mosses, these first higher plants, whose stems were already provided with vascular bundles (systems of tubes for conducting fluids), mark the most complex organizational stage hitherto in the evolution of self-nourishing organisms. Psilophytes, which appeared in the late Silurian period, can be followed almost to the end of the Devonian. The

4, 5 Two examples of psilophytes from lower Devonian strata. These were the first known primitive types of dry-land plants, usually less than a foot high.

16

Silurian period also marked the end of the age in which algae were the dominant type of plant.

Horsetails, club mosses, and ferns

In lower Devonian strata on the Old Red Continent, we find many other groups of spore-bearing vascular plants as well as psilophytes. Chief among them are the club mosses, horsetails, and—from the mid-Devonian onward—ferns. The abundance of these Devonian finds shows that the struggle waged by plants to occupy dry land since the Cambrian period had finally been won.

In mid-Devonian strata the more highly organized ferns slowly start to oust the psilophytic flora, and in upper Devonian strata they already look like trees. Horsetail and club moss types also developed at this time. Many of these plants reached huge dimensions and the accumulation in moist places of their dead remains gave rise in upper Devonian strata to the first large coal seams. During the Devonian period, the Old Red Continent thus provided plants with all the conditions enabling them to complete their millions of years' migration from ocean to dry land.

In the next (Carboniferous) period of the Paleozoic era, a mighty folded mountain range was formed in places previously covered by sea. In the innumerable lagoons, river deltas, and waterside meadows along its sides, luxuriant vegetation flourished in the warm, steamy climate. Tremendous amounts of vegetable matter collected in many of these places and, in time, by gradual chemical processes of carbonization, they were transformed to great coal deposits.

Carboniferous and Permian plants

In coal we can often find beautifully preserved remains of plants, showing us that many new types appeared during the Carboniferous period. The pteridosperms (plants resembling ferns but already bearing seeds) are particularly characteristic and important. They represent an intermediate step in evolution between ferns and cycads, palm-like plants to which the pteridosperms are most closely related. New types appeared throughout the Carboniferous period, including further gymnosperms (p. 16)—the cordaits and conifers. *Cordaites* was a genus of tall trees with leaves about a yard long. The members of this now completely extinct group contributed an important com-

6 Cordaits, the principal gymnosperm plants of the Carboniferous, reached a height of 100 feet or more, and were topped by a thick crown.

7 The giant club moss *Sigillaria* grew to a height of about 65 feet and was one of the main components of the coal forests of Europe and North America. Its cones grew directly from the trunk, just below the head.

8 *Diplopteridium*, a primitive seed-bearing fern of the lower Carboniferous, reached a height of 10 feet.

ponent to coal seams. Conifers were just starting to develop, so their forms were not yet very varied.

Among the most characteristic and familiar plants of the Carboniferous period are the giant club mosses and horsetails, which grew to veritable trees. The best known of the former are *Lepidodendron*, a giant scale-tree that reached a height of 100 feet, and *Sigillaria*, which was about 70 feet tall. The trunks of these club mosses split up into two or more branches, each of which ended in a tuft of long, narrow leaves. Among the most characteristic giant horsetails was *Calamites*. This was a tall tree with a distinctly segmented trunk and branches; and, since horsetails like moisture, they grew in swamps and marshes.

Ferns, however, were undoubtedly the most beautiful and picturesque plants in the coal forest. Fragments of their fossilized leaves and stems are to be found in every paleontological collection. The tree ferns, which grew to a height of 30 to 50 feet, were especially striking, as they had slender trunks surmounted by bright-green, feathery crowns.

Spore-bearing plants still predominated at the beginning of the Permian period (the last subdivision of the Paleozoic era), but by the end gymnosperms were in the majority. Among these we find plants that became plentiful during the Mesozoic ("middle life") era, which followed. The difference between early and late Permian vegetation is very pronounced. It marks the transition from the early period of the development of plants to the middle period, which is characterized by the predominance of gymnospermous plants.

In lower Permian strata the giant club mosses gradually disappeared, together with many seed-bearing ferns and some of the horsetails. Some new species of ferns appear, which are characteristic of the Europe of that time; and the flinty, fossilized trunks of ferns of the genus *Psaronius* are also found. Cordaits become rare in lower Permian strata, but new species of ginkgo tree make their appearance. By this time the climate was drier, so that conifers flourished. In the southern hemisphere tongued ferns of the genus *Glossopteris* and associated plants spread out over Gondwanaland. (Genus, plural genera, is from a Greek word meaning "kind" or "race".)

Animals with skeletons

In Precambrian times animals had a simply organized body without a proper skeleton. But fossil remains from the Paleozoic era show that

18

many of their successors now possessed a hard outer skeleton that shielded and supported the vulnerable parts of the body. This enabled them to build larger and more complex bodies and also gave them better protection against enemies. The development occurred in the Cambrian period (at the very outset of the Paleozoic era), which enabled the groups involved to develop very rapidly. Their many and well-preserved fossil remains have been found all over the world, in striking contrast to the extreme rarity of Precambrian finds.

Some scientists regard this sudden explosive development as proof that, at the time in question, the concentration of atmospheric oxygen had increased to such an extent that higher organisms could now develop. High up in the atmosphere a layer of ozone sufficiently thick to absorb the harmful effects of ultraviolet radiation was formed, thereby permitting life to expand into the oceans. The increase in atmospheric oxygen concentration also meant an increase in the rate of vital processes. According to one theory, animals could develop hard shells and skeletons only when the amount of energy available for the relevant metabolic processes exceeded that needed for the essential vital functions. One source of such energy was the raised oxygen concentration. Animals adapted themselves quickly to the changes in their environment and formed different types of armour and skeletons. They were very diverse, but all of them still lived in the sea—conditions were not yet suitable for animals capable of breathing atmospheric oxygen.

The early Paleozoic fauna was so varied that in it we already find representatives of practically all the major groups of invertebrates. This high degree of diversification in the Cambrian period must have involved a long previous history of evolution, although the material from the Precambrian era is so scarce that we know none of its details.

9 A leaf fragment of *Lebachia piniformis*, the earliest known representative of the conifers, commonly found in lower Permian strata.

Trilobites

Without a doubt, the most advanced animals of the Paleozoic era are the arthropods ("joint-footed"—segmented animals with an outer skeleton) known as the trilobites ("three-lobed"). We now know that 60 per cent of the known species of animals at the beginning of the Paleozoic era belonged to this group. (So far only one Precambrian fossil arthropod has ever been found, in Australia in 1964.) But by the earliest part of the Cambrian period we know that the trilobites were undergoing amazing differentiation and were splitting up into hundreds of genera and species, many of which soon died out. Numerous

19

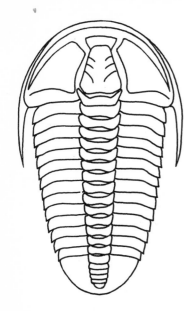

10 A typical trilobite, showing the plate-like structures on either side of the glabella and the three longitudinal lobes into which the body was divided. (Actual size 2–2³/₄ in)

trilobites still lived in the next period (the Ordovician), however, and in these strata we can see further, though less rapid, evolution and the development of new genera. They started to diminish in numbers in the Silurian period and became still rarer in the Devonian. By the Carboniferous and Permian periods only a single trilobite family (Proetidae) was left, and by the end of the Permian period its members too died out. The trilobites spread everywhere and many genera are found all over the world. They are extremely valuable for comparing and correlating strata in different continents.

During the Carboniferous period other land-dwelling arthropods also started to develop, and these included many types of insects—centipedes, scorpions, spiders, and so on. There was, for instance, a primitive dragonfly, *Meganeura*, which had a wingspan of up to 30 inches, and a centipede, *Arthropleura*, which grew to a length of five feet.

Goblet-shaped animals known as Archaeocyathinae abounded in Cambrian seas at the outset of the Paleozoic era, where they played the same role as the corals did in later periods. They lived in warm, shallow water and had a chalky skeleton. In time their skeletons accumulated to form huge limestone masses, testifying to the fact that they really existed and that in these places there were once warm, shallow seas.

Brachiopods, echinoderms, and mollusks

Brachiopods ("arm-footed"), bivalves similar in appearance to clams, also abounded from the beginning of the Paleozoic era. In Cambrian strata they form about 30 per cent of the known fauna. The shells of later Cambrian species were formed chiefly of calcium carbonate, and were hard. In some places they occurred in huge numbers and their dead remains constituted an important rock-forming element. In Paleozoic marine faunas, brachiopods are more numerous than any other animals. Hardly any marine sediments of the time are without them.

The echinoderms ("spiny skinned", such as starfishes and sea urchins) also formed an important element in Paleozoic faunas. They had simple bodies, the radially symmetrical forms of later periods having yet to be evolved. The older groups, which are already found in lower Cambrian strata, include Eocrinoidea and crinoids (sea lilies). Some primitive forms of echinoderms, such as the pouch-like cystoids, were covered with large numbers of irregularly arranged

20

plates. Some already had a stalk for attachment. In later forms the stalk was a common feature. Although the sea lilies attained their greatest development during the Carboniferous period, in a somewhat modified form they have survived through all the geological periods down to the present.

At the beginning of the Paleozoic era there were very few true Mollusca (the brachiopods mentioned above are classified as Molluscoids). They belonged to the classes Gastropoda ("belly-footed"— snails, etc) and Amphineura ("coat-of-mail shells") and, less frequently to the classes Lamellibranchiata ("plate-gilled" — bivalves) and Cephalopoda ("head-footed"—octopuses, etc). But by the middle of the era they were very abundant. Bivalves in particular were represented by members of all the main known orders, and the evolution of cephalopods was likewise in full swing. Freshwater species of bivalves are found in strata as early as the Devonian, and are common in the Carboniferous and Permian. In later Paleozoic rocks we also find bivalves that lived in fresh and brackish water. Gastropods too abounded during the Paleozoic era, but the first freshwater forms did not appear until the Carboniferous period. The commonest cephalopods were the nautiloids (represented today by the living *Nautilus*), whose evolution culminated in the Silurian period. Toward the end of the Paleozoic they were for the most part replaced by the ammonites. These are cephalopods, typically with spirally coiled, many-chambered shells, often radially ridged like a ram's horn. (The name comes from the "horn of Ammon"— Amun, king of gods in ancient Egypt, was represented as ram-headed.) Chief among the ammonites was the group known as goniatites, which flourished in the Devonian and Carboniferous periods. They are immensely important for the correlation of different marine strata.

Two more groups—the graptolites and the coelenterates—merit special attention. Graptolites ("written stones", often preserved looking like pencil marks in the rock), were marine animals living in colonies; their remains help to determine the age of different layers. Coelenterates ("hollow inside") were hollow-bodied animals, among which the corals acquired particular importance. The corals were represented by wrinkled forms (Rugosa) and by flat forms (Tabulata). Silurian corals were important contributors to limestones of mainly organic origin. Another widespread group occurring in the middle periods of the Paleozoic era were the stromatopores, which were related to the corals. These organisms, whose origin is still a matter of debate, secreted low-spreading, hard, limy skeletons, some of which measured over six feet across. They form an important

11 Graptolites consisted of a colony of tiny marine organisms, each occupying a separate "tooth" in a colony made up of one or more branches. They were apparently attached to drifting seaweed or wood. (Natural size)

21

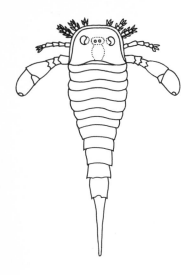

12 A Silurian eurypterid, an early marine relative of the scorpions and spiders. (Actual size 6 in)

structural component of Silurian and Devonian limestones. Another group of coelenterates very plentiful in the Paleozoic era were the conularians. Relatives of the jellyfish, they appeared in the Ordovician period, started to die out in the Devonian, and finally disappeared at the beginning of the Mesozoic era.

The biggest of the Paleozoic mobile invertebrates (animals without backbones) was *Eurypterus*, a huge arthropod with pincers. Eurypterids (the word means "broad-flippered"), which are sometimes classified midway between the trilobites and the scorpions, appeared in the Cambrian period, attained the peak of their development halfway through the Paleozoic, and have been slowly dying out ever since. Some Paleozoic forms were really giants. In the Silurian and Devonian periods they often measured 10 feet long and were among the largest animals then in existence.

The end of the Paleozoic was marked by widespread development of the Foraminifera ("hole-bearers")—mostly single-celled, tiny creatures encased in a chambered, perforated shell. Some of their genera, especially in the Carboniferous and Permian periods, were, however, very large and occurred in such quantities that their remains became important rock-forming materials.

Toward the end of the Paleozoic the evolution of the foraminiferans continued, but a whole series of other groups had started to decline. The trilobites were finally dying out, the four-horned corals were disappearing, and the significance of the brachiopods was on the wane. Bivalves were relatively abundant and their many genera were already closely related to Mesozoic types. Cephalopods, among which the first true ammonites now appeared, also continued to play an important part.

Animals with backbones

The earliest remains of vertebrates (backboned animals) appear in Ordovician sediments. They are parts of the armour of primitive, jawless, fish-like vertebrates of the ostracoderm ("shell-skinned") group and have been found in lower Ordovician strata in Estonia (U.S.S.R.) and in middle Ordovician layers in the United States. From them we can follow the evolution of higher and more complex organized groups, that is, the first jawed fishes, known as placoderms ("plate-skinned"), although these did not develop until the Devonian period. They included strange, heavily armoured types such as *Pterichthyodes*. At the beginning of the Devonian, placoderms were small

22

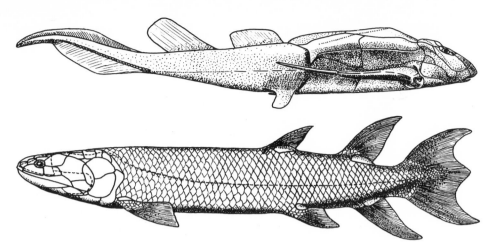

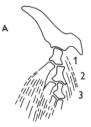

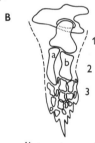

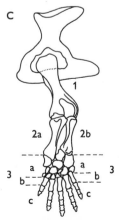

13 One of the first jawed fishes (*Bothriolepis*), a placoderm, showing clearly the "plates" of bony armour from which they get their name. (Actual size 1 ft 3½ in)

14 *Eusthenopteron*, a Devonian crossopterygian (fringe-finned fish), one of the ancestors of land quadrupeds. (Actual size 1 ft 9½ in)

15 Diagram showing how the bones of an amphibian hindlimb (C) may have developed from those of a crossopterygian fin (A) through an intermediate (imaginary) stage (B).

A: One side of the pelvis and fin skeleton of a crossopterygian (see fig. 14). The first segment is formed by a single bone — the future thighbone (1). The second and third segments of the limb potentially represent the shin-bone (2) and the foot (3).

B: Intermediate stage. Even the second segment of the limb has developed into two bones (2a, 2b) corresponding with the tibia and fibula. The future skeleton of the foot is represented by tiny bones of the third segment (3).

and resembled ostracoderms. But they soon increased in size, some of them to gigantic proportions, like *Duncleosteus*, which measured about 35 feet. These carnivorous monsters must have terrorized Devonian seas. Alongside the placoderms the first ancestors of the true sharks appeared in the middle of the Paleozoic era; during the late Paleozoic some of them even lived in fresh water.

Various groups of true bony fishes (Osteichthyes) also started to evolve, and at the end of the Devonian the first amphibians (Ichthyostegalia) developed from one of them. But practically all the groups of primitive Devonian fishes died out by the end of this period. The only exceptions were the Acanthodii, a curious group of spiny fishes, which survived into the Permian, and a few groups of bony fishes. The ancestors of these bony fishes appeared relatively early, at the beginning of the Devonian.

The bony fishes spread very rapidly and by the end of the Devonian they formed the main group of freshwater vertebrates. Almost from the outset of their evolution they split up into two different groups. The first is still very large and includes about 90 per cent of all existing species of fish. This group is termed Actinopterygii ("ray-finned") from the characteristic structure of their fins. The other group of bony fishes is now very small. But from the aspect of evolution it is very important, as it was the basic vertebrate group from which all further land-dwelling groups, and finally man himself, developed. This group has two sub-groups: Dipnoi (lungfishes) and Crossopterygii (fringe-finned fishes). It was from the last of these that the direct ancestors of quadruped dry-land vertebrates developed.

C: Pelvis and hindlimb skeleton of a primitive quadruped. The thighbone articulates at the knee joint with two bones — tibia, or shin bone (2a) and fibula (2b) — as in all quadrupeds. The bones of the third segment are now organized into tarsus (3a); metatarsus, or foot-arch (3b); and phalanges, or toes (3c), of which there are primitively five in all land vertebrates.

23

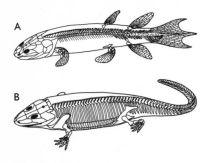

16 Comparison of a crossopterygian fish of the late Carboniferous (A) with an amphibian of the same period (B), showing general similarities but skeletal differences. Notice the bones of the fish's fin skeleton, which represent an early stage in the evolution of a limb for an air-breathing land quadruped (see fig. 15).

Crossopterygians, which had brush-like fins attached to stumpy "limbs", lived in fresh water and flourished during the Devonian. In subsequent geological periods they gradually died out, however, and today they have only a single representative, the genus *Latimeria*, the coelacanth, which lives in deep water off South Africa (p. 30). The first land-dwelling vertebrates, whose remains were found by a Danish expedition in the 1930s on the island of Imer, east of Greenland, evolved from crossopterygians during the Devonian.

The transition from water to dry land was an epoch-making event in evolution. Naturally it took a long time, because various physiological and structural modifications had first to be made. Vertebrates that started to leave the water moved over swampy ground. They still used swimming motions and propelled themselves forward with rapid twists and turns. It was not until later that they developed the use of their limbs. As soon as Devonian crossopterygians were able to live permanently on dry land, four-footed animals began to evolve from some of their groups. These were the first amphibians, known as stegocephalians ("roof-skulled").

Many amphibian forms evolved during the Carboniferous, and to some extent during the Permian period, when they occupied fresh waters and moist parts of the land. They varied in size from a few inches long to the proportions of a small alligator. Their eggs and larval stages (tadpoles) still required water for their development.

The first reptiles

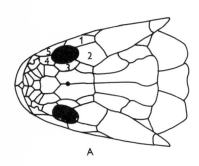

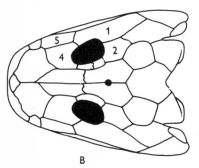

With droughts occurring and lasting for increasing periods of time, far-reaching climatic changes characterized the end of the Carboniferous. This led to the development of another major group of animals —the reptiles, which started to evolve from the stegocephalians. Unlike amphibians, which fertilized their eggs in water, reptiles were able to fertilize their eggs inside their bodies and then to lay them complete with a supply of food (the yolk) protected by some sort of coat (the shell). Although their origins go back to the late Carboniferous, their rapid development did not really start until the Permian. From then on we can follow several main evolutionary lines that persisted into the Mesozoic era.

Later Carboniferous reptiles were still very primitive. More abundant at this time were the relatively large pelycosaurians ("pelvis

17 Dorsal view of the bones in the skull of a crossopterygian fish (A) and those of a primitive amphibian (B). The general similarity is striking though details and proportions differ. Note, for example, the correspondingly numbered bones surrounding the eye socket, and the parietal foramen, a small opening in the mid-line for the vestigial third eye.

24

lizards"), which are also described as "mammalian" reptiles, as an evolutionary line leads straight from them to the Mesozoic mammals.

The Permian period marked the end of the Paleozoic era, which lasted some 345 million years. During this time life developed rapidly and gradually moved from the sea to the land, so that by the end of the era we find types capable of existing and surviving in distinctly dry and inhospitable surroundings. Among these we already find forms that gave rise to lines, from which mammals, and hence man himself, eventually evolved.

Mesozoic era

According to the old geologists the Mesozoic ("middle life") era is the second era in the earth's history. It forms a kind of transitional period, both in the development of the earth's crust and in the evolution of life. It starts with the end of the great Hercynian mountain-folding processes and ends with the beginning of new and greater Alpine mountain-forming events. In the southern hemisphere Gondwanaland broke up, but otherwise the Mesozoic was relatively quiet with only small and inconstant continent-forming movements. It lasted about 160 million years.

Geologists divide the Mesozoic into three periods—Triassic, Jurassic and Cretaceous, the first two being considerably shorter than the third. The Cretaceous period is so long (71 million years) that it is often divided into two distinct phases, the lower and the upper. The dividing line between them is an important landmark in the evolution of living organisms.

In the development of life the Mesozoic is a period of transition from old, more primitive types to new, more advanced types. No four-horned coral and no trilobite or graptolite overstepped the boundary between the Paleozoic and the Mesozoic eras. The Mesozoic world was very different from that which preceded it, and so its fauna and flora also differed.

More advanced plants

The more advanced gymnosperms ("naked seed" plants) had already begun to assert themselves in the late Permian period. The age of the predominance of algae, psilophytes, and cryptogamous (spore-bearing) vascular plants belongs to the early era of botany, while

18 A cycad of the Jurassic period, showing the low, barrel-shaped trunk typical of these gymnosperms, which flourished in the Mesozoic era. (Actual size 3 feet)

the development of the higher gymnosperms constitutes the middle era, which lasted until the beginning of the late Cretaceous period. After that the first flowering plants with a seedcase appeared, and with them dawned the modern age of botany.

The development of gymnosperms was an event of primary importance in plant evolution. To be able to reproduce, the old spore-bearing plants of the Paleozoic era had to have water, or at least a humid environment. This was a great obstacle to their spreading. The advent of seeds meant that plants were no longer confined to the water. They could now be fertilized without water, for instance, by insects. Furthermore, whereas the spore is a single cell with a relatively small amount of nutrients, the seed is a structure of many cells containing a far larger supply of nutrients. It thus provides much better conditions for the embryo to develop. Seeds also have outer cases to protect the embryo from external damage. These improvements all provided better chances of survival and further evolutionary progress for the species. The ovules (female sex cells) of the first seed-bearing plants developed in the open, uncovered, on special leaves, and the seeds had no outer casing. Plants with naked ovules and seeds are therefore called gymnosperms.

Among the commonest and most striking gymnosperms at the beginning of the Mesozoic era were the cycads. They had either tall trunks like trees, or low, thick, barrel-like stems, with long, tough and usually feathery leaves (for example, *Pterophyllum*, "wing leaf"). In appearance they closely resembled tree ferns or palms. From the evolutionary aspect one of the most important families of cycads was the Bennettites, which grew as bushes or trees. In general they were similar to the original cycads, but they had already begun to develop seedcases and other features that were the outcome of their adaptation to a drier climate.

New types appeared in the Triassic period. Conifers, including groups like firs, cypresses, yews, and others, were abundant. Among members of the ginkgo family we find the genus *Baiera*, which had deeply cleft, ragged leaves. Ferns flourished in damp, shady spots and beside water. Creeping ferns took root on rocky ground, and horsetails grew in bogs, although no longer to the same height as their forebears.

Plant development reached its peak during the Mesozoic era. The warm, tropical climate of places that are today located in the temperate zone was ideal for the growth of tree ferns, while smaller ferns and herbaceous types of plants grew better in cooler conditions. But gymnosperms, including cycads, were still dominant.

26

Important changes took place in plant evolution during the Cretaceous period. The lower Cretaceous flora still closely resembled that of the Jurassic. Gymnosperms (mainly cycads) were commonest, but conifers, ginkgoes, and seed ferns also abounded.

Flowering plants

Angiosperms marked a higher evolutionary stage in plant development. Their seeds are enclosed in cases and they have specially developed sex organs (stamens and a pistil), which are usually surrounded by brightly coloured petals and a green calyx. These flowering plants probably originated at a time before the Cretaceous period, possibly in cold conditions at high altitudes. As the climate gradually became cooler during the early Cretaceous, they quickly spread downward and upward. They soon became adapted to their new environment, and after that they developed at an astonishing rate.

19 A spray from an angiosperm of the upper Jurassic, showing fertile flower heads.

The remains of plants that were definitively angiosperms are found in lower Cretaceous strata of Europe and Asia. They occur over a wide area in deposits of more or less the same age and already display remarkable diversity. From then on angiosperms evidently gained the upper hand and by the beginning of the upper Cretaceous their supremacy was complete. Cretaceous angiosperms were evergreen, tropical, or subtropical types, such as eucalyptuses, magnolias, sassafrases, tulip trees, japonicas, cinnamons, bay trees, walnut and plane trees. In higher altitudes these warmth-loving plants were accompanied by types suited to a more temperate climate, such as oaks, beeches, willows, and birches. Among the gymnosperms conifers, such as sequoias and pines, were numerous.

By now gymnosperms were starting to decline. Some of them still persisted, but their numbers constantly diminished. The only gymnosperms remaining in large number were conifers, many species of which have survived down to the present.

Plants made tremendous progress during the Mesozoic era, and in the rapidity of their evolution they actually outstripped the animals.

Cephalopods

The Mesozoic invertebrates were already more modern in character. Among the most typical and important were the cephalopods (relations of the modern octopus and cuttlefish). In particular these included the ammonites, which had four gills and a spiral outer

20 The leaf of a sweet gum, a broad-leaved tree that grew in the forests of northern Europe in Tertiary times. Related species of the present survive only in North America and Asia Minor.

27

shell, and the belemnites, which had two gills and a spear-shaped inner shell covered by a soft, fleshy body. Ammonites were so abundant in the Mesozoic that in places their shells form a major component of seabed sediments. In fact the Mesozoic is often called the Age of the Ammonites. Their development started in the Triassic, at the beginning of the Mesozoic, during which over 400 new genera evolved. Members of the order Ceratitida were the most characteristic. Many were quite small, with shells an inch or so in diameter, while there were some giants of two feet and more. Belemnites, which look like spearheads, are sometimes called "devil's thunderbolts". The name comes from the Greek word for a dart.

Most of the older types of ammonites died out at the end of the Triassic, but the families Phylloceratidae and Lytoceratidae still lived in Jurassic seas, where they evolved so explosively that they soon made up for previous losses. Both ammonites and belemnites still abound in Cretaceous strata, but their numbers start to decrease in the upper layers. Among the ammonites we find forms with only partly twisted shells *(Scaphites)*, completely straight shells *(Baculites)*, or irregularly spiral shells *(Nipponites)*. These oddities are probably a sign of disorganized individual development and of over-specialization.

The two-gilled cephalopods are of great significance in Mesozoic rocks, because the fossils of some of their genera (for instance, *Actinocamax* and *Belemnitella)* preserved in upper Cretaceous strata, serve as "zone fossils" and help to establish the dates of the layers in which they are found. By the end of the Mesozoic, however, all the ammonites and most of the belemnites had died out. The latest forms of some evolutionary lines of ammonites are often characterized by abnormally large shells. Members of the genus *Parapachydiscus*, for instance, had a twisted shell averaging eight feet in length. Only a few groups of nautiloids survived into the Tertiary.

Corals and mollusks

The four-horned and flat corals did not survive into the Mesozoic era. Their place was taken by the six-horned corals, which multiplied enormously during the Triassic period, when their communities formed huge reefs, which still exist today. Some groups of brachiopods also developed slightly during the Mesozoic, but most of them declined. Mesozoic echinoderms were represented by several types of crinoids (sea lilies), which were quite abundant in the Jurassic period and part of the Cretaceous in some areas of the shallow seas. But their evolution was nothing compared with that of the sea urchins (Echinoidea), which developed innumerable species. The number of serpent stars and starfishes also increased.

Mollusks developed far more than during the Paleozoic era, and already in the Triassic they had many characteristic genera. At the beginning of this period we also find the first oysters, which became one of the largest mollusk groups in Mesozoic seas. New groups of mollusks continued to evolve in the Jurassic, when two of the most familiar genera were *Trigonia* and *Gryphaea*. In Cretaceous formations we find curious types of bivalves with cornet-shaped shells fitted with a lid. These animals made an important contribution to the formation of chalk cliffs. The most characteristic bivalve of the Cretaceous was the genus *Inoceramus*, some of whose species were about two feet long.

In some places we find large quantities of gastropods (snail-like univalves). In the Jurassic period there was a new burst of development among the foraminiferans (p. 22), which continued through the Cretaceous and into recent times. Protozoans (single-celled animals) were important rock-forming elements in the Mesozoic and are of great significance for determining and comparing the age of different layers. Different types of sponges and some groups of arthropods, particularly insects and ten-footed crustaceans, also appeared in the Cretaceous.

Fishes

The Mesozoic era is marked by tremendous expansion of many groups of vertebrates. Of the Paleozoic fishes only a few survived into the Mesozoic, such as the genus *Pleuracanthus*, the last representative of Paleozoic freshwater sharks, known from Triassic freshwater sediments in Australia. Marine sharks developed considerably during the Mesozoic and most of their still extant families were already represented in Cretaceous seas.

The ray-finned fishes lived in fresh water, but in the Triassic period they also invaded the seas, where they multiplied tremendously and have been the most prominent group ever since. We have already met (p. 23) the Paleozoic group of ray-finned fishes from which the first land-dwelling vertebrates evolved. The typical genera of this group died out during the Mesozoic, although in Cretaceous strata we still find some remaining representatives. Up to 1938 these were the last known members of this group, which paleontologists supposed

29

to have been extinct since the Cretaceous. But in that year an event occurred that made all paleontologists sit up. A creature unknown as a living animal was caught off the coast of South Africa. Scientists who examined this curious fish confirmed that it belonged to the supposedly extinct group of the coelacanths ("hollow-spined" fishes) and that it was the only one known to have survived to the present day. They named it *Latimeria chalumnae*. Such survivors are called "living fossils".

Amphibians

Stegocephalians (primitive armoured amphibians) were still quite plentiful in some places in the Triassic period, but more specialized types soon appeared. The stegocephalian *Trematosaurus*, the only amphibian to prey on sea fish, was also a Triassic animal. Stegocephalians began to die out by the end of the Triassic, but the predecessors of the present-day frogs started to evolve, probably from Paleozoic stegocephalians. Their Triassic representative is the genus *Triadobatrachus*, of which only a single, incomplete skeleton has so far been found, in the north of Madagascar. Representatives of the early frogs were much more numerous by the Jurassic period, however. Frogs developed more rapidly in the Cretaceous, but their evolution culminated in the Tertiary era (p. 36) and in recent times. The first relations of modern newts and salamanders also appeared in the Jurassic, although their real development did not start until the Tertiary.

Reptiles rule the world

21 Diagram showing the explosive development of reptiles in the Mesozoic, and related evolutionary lines of other major vertebrate classes.

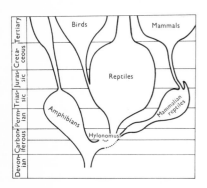

The class of animals that underwent the greatest development during the Mesozoic was that of the reptiles, which dominated the world. New genera and species of such diversity evolved, often of such huge proportions that some of them were the biggest and strangest land animals in the earth's history. In body structure, as we have seen, the oldest reptiles were closely related to the stegocephalians, but a series of independent evolutionary lines soon developed from these primitive forms. Among these were the cotylosaurs ("cup-jointed lizards"), small or moderately large, ungainly, omnivorous reptiles that died out at the end of the Triassic period. But a number of new evolutionary lines developed from their descendants. Another primitive group were the pelycosaurians ("mammalian reptiles"), which appeared at the end of the Carboniferous period (p. 24). The

30

New World population of these animals became extinct halfway through the Permian period, but their Old World relatives developed and expanded, forming a number of new branches that are grouped in a separate order, Therapsida. The carnivorous line of this suborder, Theriodontia, greatly resembled primitive mammals, and it was actually from some of its families that the first mammals evolved at the end of the Triassic.

Other important groups of reptiles also developed during the Triassic period. They included testudinates (turtles and tortoises), the highly specialized ichthyosaurs (fish-like reptiles), the placodonts (armoured reptiles with a functioning third eye), and others. A group particularly important from the evolutionary aspect are the thecodonts (small, carnivorous reptiles with socketed teeth). It was mainly from them that there eventually evolved the many groups of running and flying reptiles, so characteristic of the Mesozoic era, and also groups well known today, such as crocodiles and birds.

Everybody has heard of the dinosaurs. These reptiles evolved from a thecodontic ancestor in the Triassic period, and during the Mesozoic era they became the dominant animal group. The Jurassic and Cretaceous periods were the heyday of the dinosaurs. In the Jurassic we find among them veritable monsters up to 90 feet from nose to tail and weighing as much as 50 tons. *Brontosaurus, Diplodocus,* and *Brachiosaurus* are some of the most striking. The evolution of the giant saurians continued during the Cretaceous. The well-known iguanodonts lived in Europe, while America was the home of horned dinosaurs resembling rhinoceroses *(Triceratops, Styracosaurus)* and of the heavily armoured ankylosaurians. Herbivorous dinosaurs included many strange forms, such as *Anatosaurus,* which had jaws like a duck's bill, and *Trachodon,* which resembled a huge plucked ostrich. Carnivorous types also flourished, among them *Tyrannosaurus, Gorgosaurus,* and *Tarbosaurus,* some of the biggest beasts of prey of all time.

Many families of aquatic reptiles also evolved during the Jurassic period. The plesiosaurs ("near-to-a-lizard") were common and voracious beasts of prey in Jurassic and lower Cretaceous seas. They had a small head, a long, thin neck, a turtle-like body, flippers for limbs, and a short tail.

At this time the first crocodiles *(Steneosaurus)* appeared, together with winged reptiles (Pterosauria). The latter probably evolved from thecodontic forebears at some time during the Triassic period. Their best known genera are *Rhamphorhynchus, Pterodactylus* and, somewhat later, *Pteranodon.* Winged reptiles became extinct at the end of the Cretaceous, leaving no descendants.

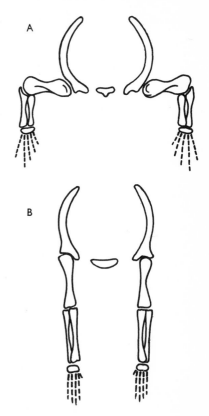

22 Axial view of the relative positions of the bony girdles and supporting limbs in a primitive reptile (A) and a mammal (B). In the former the points of support provided by the feet are widely offset from the load of the body; in the latter they are vertically beneath it—a much more efficient arrangement.

31

Ancestors of birds

The first representatives of the birds appear in Jurassic strata. *Archaeopteryx*, the earliest specimen we know, was found near Solnhofen in Bavaria. Their evolution progressed rapidly during the Cretaceous period, when the characteristic genera were *Ichthyornis* and *Hesperornis*.

Cretaceous seas were also inhabited by carnivorous saurians resembling monitor lizards except for their paddle-like limbs. At the end of the Cretaceous the first snakes (Ophidia) made their appearance.

First mammals

The first insignificant, tiny, mouse-sized mammals evolved from "mammalian" reptiles at the end of the Triassic. They developed and diversified modestly throughout the Mesozoic, most of the early orders becoming extinct before its end. Their descendants, the first marsupials (animals with pouches) and insectivores, first appeared in the late Cretaceous. These two groups continued their evolution through the Tertiary era and have survived down to the present.

At the end of the Cretaceous period, practically all of its characteristic groups of reptiles (dinosaurs, ichthyosaurs, plesiosaurs, pterosaurs, marine saurians, and many others) became extinct. Tremendous mountain-forming processes and continental movements caused extensive geographical and climatic changes. Most of the old Mesozoic world and its inhabitants altered, died, and disappeared; from the wreckage there grew up a new Tertiary world, in which life entered on fresh stages of evolution and came to be more as we know it today.

Cenozoic era

The last phase in the development of the earth is known as the Cenozoic ("recent life") era. It lasted about 65 million years and is of especial importance to us, as it was during this time that the group of primates evolved, from which man himself finally developed. The mighty Alpine mountain-folding upheavals died down and gradually the earth's surface acquired its present form. Geologists usually divide the Cenozoic era into two geological sub-eras of unequal length—Tertiary (third) and Quaternary (fourth). The Tertiary is by far the greater, and the Quaternary is distinguished only because it constitutes a clearly defined geological age, with some unique features, such as ice ages, and the interglacials dividing them (p. 39).

32

The length of the Tertiary is estimated at about 65 million years; geologists divide it into five periods, the Paleocene, Eocene, Oligocene, Miocene, and Pliocene. Like most of the others, this era also started with marked movements of the earth's crust. Extensive areas of some parts of the continents sank, so that the sea invaded Europe as far as the Russian plateau and swallowed up the outlying parts of both the Americas and large areas of Africa. At the end of the Oligocene, new continental lifts occurred, so that the coastlines altered once again. During the Miocene new mountain ranges and chains were formed. The Alps, the Pyrenees, the Carpathians, and the Himalayas acquired their present form. The end of the Tertiary is not sharply defined in sedimentary rocks. It was characterized chiefly by a rapid worsening of the climate, a drop in temperature, and the gradual formation of an ice sheet.

Tertiary plants

The Tertiary flora, which began to develop at the end of the Cretaceous period, has many features in common with present-day vegetation. The evolution of angiospermous plants, including both monocotyledons and dicotyledons (plants with one seed-leaf or two), was in full swing. Conifers were also very widespread, although the number of their species was not large. Species that now grow in the warmer regions were represented among them, showing that the climate was tropical to subtropical, and humid. In the warm, damp climate that developed after a brief cooling interlude at the beginning of the Tertiary, the subtropical flora spread far to the north. Even the polar regions were warm, with the result that magnolias, bay trees, chestnuts, and other warmth-loving plants grew in Greenland and Spitsbergen.

In the Oligocene, japonicas, cinnamon and camphor trees, fig trees, plane trees, palms, and so on, were plentiful in north and central Europe. Giant yews and enormous sequoias grew in swampy regions. Lignite (soft coal) is formed largely from the trunks of these trees. In more temperate regions like the Baltic, other types of conifers, such as pines *(Pinus succinifera)*, flourished. The amber found on the shores of the Baltic is actually the fossilized resin of these trees. In the Pliocene the climate started to change. At first it was still suitable for the warmth-loving flora, and palms, magnolias, and cinnamon trees grew in western and central Europe. The fossilized remains of a yew tree of a type that still grows today in Florida was actually found at latitude 80°N. In time, however, the weather grew steadily colder, so that palms and other plants requiring warmth retreated slowly but surely southward.

33

In central Europe the climate remained mainly warm and damp, as we can see from the extent of the spreading deciduous forests. These contained large numbers of elms, oaks, plane trees, sycamores, chestnuts, and other trees that today grow wild in southern Europe and south of the Caucasus. In more northerly areas coniferous forests composed chiefly of firs and yews predominated. In the earlier (and warmer) part of the Tertiary sequoias, yews, and many types of spruces still abounded in central Europe. Two genera of ginkgoes still survived in the Tertiary era, but one of them later became extinct. The climate grew worse throughout the whole of the later Tertiary, however; by the end of it some of the trees in deciduous and coniferous forests required a milder climate, and so retreated from temperate Europe farther south, or died out altogether.

Tertiary animals

Such far-reaching geographical changes occurred in most parts of the world at the end of the Mesozoic and the beginning of the Tertiary era that many groups of animals became extinct. These included the belemnites, the ammonites, and many other groups of invertebrates. Most of the older types of fishes and many Mesozoic genera of amphibians disappeared. The dinosaurs and most of the other reptiles likewise vanished from the earth. Groups that survived the crucial ordeal, such as tortoises, crocodiles, snakes, and lizards, are still with us today, though in greatly reduced numbers.

Many new genera that have survived down to the present appeared among the invertebrates at the beginning of the Tertiary. Of the protozoans, the foraminiferans and radiolarians (microscopic animals with hard parts) reached the peak of their development. The brachiopods now rapidly declined and only isolated genera survived into the Quaternary. Mollusks, however, went on developing. Some of the commonest bivalves of the Tertiary, such as the true oysters *(Ostrea)*, the scallops *(Pecten)*, and the thorny oysters *(Spondylus)*, are still abundant today. The gastropods underwent very rapid and explosive evolution, and continued to develop in the Quaternary. Most of these genera still inhabit present-day seas. Among the best known are *Cerithium*, the mitre shells *(Mitra)*, the spindle shells *(Fusus)*, and the conches *(Strombus)*.

34

Great changes occurred among the cephalopods. As we have seen, the ammonites completely vanished from the seas. Their place was partly taken by some genera of the nautiloids, which persisted into the Quaternary, though in far smaller numbers. The two-gilled cephalopods (Dibranchia) proved to be much tougher and more adaptable. As the groups that had survived into the Tertiary developed more rapidly, their many forms and individuals soon filled in the gaps left by types that had become extinct. The cuttlefish (Sepioidea), the octopuses (Octobrachia), and the giant squids (Teuthida)—orders that still inhabit the sea—spread particularly fast. As certain branches evolved, the size of the animals' shells became much smaller; this is true especially of the octopus.

Arthropods evolved and flourished to a high degree and many new genera appeared. The shells of ostracods (small crustaceans with a bivalve shell, living in fresh, brackish, or salt water) are an important component of the relevant sediments. The sea swarmed with crabs and lobsters. As flowering plants developed on dry land, vast quantities of every conceivable type of insect appeared. The remains of many species have been preserved for us, trapped in amber, or sealed in sedimentary strata.

Echinoderms were very abundant. This applied mainly to the sea urchins (Echinoidea), among which genera with irregular, flattened shells, or shells bulging on one side, predominated. The crinoids were declining, but they still formed a characteristic part of the marine fauna. Those that were mounted on a stalk and were very common in Mesozoic seas retired to the deeper parts of the oceans in the Tertiary. However, one group, the comatulids, continued their development. These sea lilies partly transformed their reduced stalk into a float and swam about freely in the sea. Some 90 different genera of these crinoids still exist. They are the principal representatives of a group that was once so large that its innumerable species formed huge meadows on the sea bed and provided the main material that went into the making of crinoid limestones.

Fishes and amphibians

The Tertiary era was an age of tremendous development for the bony fishes (Osteichthyes). Both tailed and tailless amphibians flourished. The most famous representative of the tailed amphibians was the large Miocene salamander known as *Andrias scheuchzeri*, which lived in Europe in the middle of the Tertiary. One of its relatives, the genus

35

Megalobatrachus, still exists in China and Japan. Frogs were fairly abundant in lakes and swamps. Their appearance was largely associated with volcanic activity. Upheavals of the earth's crust produced many new water reservoirs, such as swamps and lakes, which encouraged the development of amphibians. Huge numbers of them perished in the frequent volcanic catastrophes, but their well-preserved skeletons give us a good idea of what they looked like. In early Tertiary Europe the now extinct family of Palaeobatrachidae (frogs that lived permanently in water) predominated. Representatives of the well-known, present-day families appeared later in great numbers.

Reptiles and birds

Most of the reptiles died out at the end of the Mesozoic era, with only a few persisting into the Tertiary. The evolution of some groups of snakes continued in the early, warmer part, while the other groups of reptiles—chelonians (tortoises and turtles), crocodiles, and lizards—formed, as now, only a minor part of the animal life. Birds developed very rapidly, however. There were many more of them, both species and individuals, and their body structure was more advanced than at the end of the Mesozoic. Their evolution was helped by the large-scale spread of angiospermous plants and insects. Insects and their larvae, together with the buds, seeds, and flowers of plants, provided birds in plenty with the kinds of food that they needed. Birds, too, had virtually no enemies at that time.

Golden age of mammals

Mammals underwent explosive development during the Tertiary era. The primitive egg-laying mammals (Prototheria) and marsupials, some of whose groups survived from the Mesozoic to the Tertiary, retreated into obscurity, and from the end of the Mesozoic placental mammals (mammals that are born fully developed) began to supersede them. At first they were represented only by a few primitive insectivores, but these, or very similar forms, gave rise to later orders of placental mammals. The disappearance of the reptiles left vast areas of land and many environmental niches unoccupied. But they were soon invaded by more or less unspecialized, though highly adaptable, mammals, which spread tremendously from the very outset of the Tertiary. Since the climate was still good, the abundant vegetation provided them with huge stocks of food, of which they took full advantage. In the first period of the Tertiary (the Paleocene) we already find over 400 species of mammals. Their development then proceeded very rapidly, and by the end of the Tertiary forms appeared from which man's predecessors evolved.

We have already mentioned (p. 32) the primitive insectivores as one of the most important groups of mammals in the early Tertiary. They were not very different, on the one hand, from their Cretaceous ancestors or, on the other, from their modern descendants. Indeed, their remains are so similar to those of the first primates that even experts have difficulty in distinguishing between the two groups. This shows that they must both have originated from the same stock. At the time when the development of the insectivores stopped, however, the evolution of the primates was only just beginning.

The first rodents appeared at the outset of the Tertiary. Found almost everywhere, they formed a group of very successful mammals, although some of their families, for example, the voles, did not reach their peak until the Quaternary era.

The first ungulates (mammals with hooves) also appeared at this time. They probably evolved from the extinct group of the Condylarthra, and their oldest representatives were generalized omnivores with some characteristics that we associate rather with the carnivores than with hoofed animals. They still have a close relative, the aardvark, living in South Africa.

The many families of ungulates soon separated into two distinct groups—the "odd-toed" and the "even-toed". Odd-toed ungulates (Perissodactyla) have three toes, the central one being the strongest; they include large herbivores, such as rhinoceroses, tapirs, and horses. The evolution of the horse, which took place mainly in the Tertiary era and was rather complicated, is now comparatively well known. The process was only finally completed in the Quaternary.

Even-toed ungulates (Artiodactyla) have two equally developed major toes, sometimes two more; they include pigs, camels, and ruminants (animals such as cattle and sheep, which chew the cud). Some of these groups, especially the horned cattle (Bovidae), did not really start to evolve until the end of the Tertiary, and their further development continued through the Quaternary.

The proboscids (animals with trunks), that is, mastodons, mammoths, and elephants, were a large and important group in the Tertiary. They represent one of the most interesting stages in the evolution of the mammals. They probably originated in Africa, where their earliest known ancestor, *Moeritherium*, which was about the size of

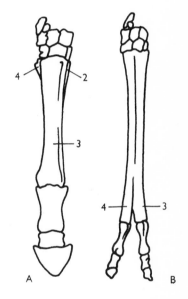

23 Comparison of skeletons of the right fore-extremity (corresponding to the wrist and hand in man) of a horse (A) and a camel (B). The horse, an odd-toed ungulate, has the main, third, digit (3) in the axis of the limb, with vestiges of the second (2) and fourth (4) flanking it. The camel, an even-toed ungulate, has two equally developed bones (3 and 4) fused into one for most of their length, with equally developed toes forming the "cloven" hoof.

37

a pig, was found. In time, they spread to the whole of Europe and Asia and eventually to both North and South America. They were a varied group in the later Tertiary, and many of their often huge bones have been found as fossils. The descendants of Tertiary proboscids persisted through the Quaternary as true elephants even in the cold parts of Europe, Asia and America, where whole carcasses, tusks, and bones have been found preserved in permafrost soils. These animals were the almost legendary woolly mammoths. Today only two genera of elephants survive, one in Africa, the other in India.

In the early Tertiary we encounter the oldest, still very primitive carnivores (Creodonta), which were later succeeded by more modern types belonging to the order Fissipedia ("separate toes"). Some groups of Tertiary cats (Felidae) displayed a tendency to rapid specialization, resulting in the famous "sabre-toothed" forms. These disappeared early in the Quaternary. Primitive man was already familiar with some of the now extinct beasts of prey, such as the cave lion, which died out only in the last ice age.

The first primates

The first primates evolved in the early Tertiary, from insectivores. These lemur-like animals mark the beginning of an evolutionary line that can be clearly followed right through to the apes (Anthropoidea), among which three different groups developed. These were the New World, flat-nosed monkeys (Platyrrhina), which mainly inhabit South America and have been separated from other primates since the Eocene. They had nothing to do with the evolution of man. The Old World, narrow-nosed monkeys (Catarrhina), ultimately from the same common stem, eventually gave rise to the third group, Hominoidea. This is a branch that separated from the catarrhine link some time toward the end of the Oligocene and formed a new group, which includes the more highly organized primates, that is, the anthropoid apes and man. With the appearance of *Homo sapiens* in the Quaternary the evolution of the primates, which took 65 million years, reached its climax.

Quaternary era

The last and shortest era in the earth's history is the Quaternary, which began only two million years ago. Geologists divide it into two periods—the Pleistocene ("mostly recent"), and the Holocene ("entirely recent"), which has lasted only 10,000 years.

Developments in the Quaternary were rather different from those in most of the other geological periods, for an increasingly cold climate left its mark on the surface of the earth and on all forms of life.

The cooling that had been going on since the middle of the Tertiary now became acute. As the temperatures fell in higher latitudes, snow and ice forming in the winter did not melt in the colder summers, so that over the years the layers of snow built up in mountainous areas and gradually became a continuous ice sheet. Under great pressure the ice flowed outward from such centres, so that in time large areas of the earth, in both northern and southern hemispheres, became covered in ice. At their maximum the ice sheets covered almost 20 million square miles of territory. In Europe they reached to southern England (as far south as the Thames), Holland, the Harz and Carpathian mountains, and the valleys of the Don and Dnieper rivers in southern Russia. In North America they extended as far south as latitude 40°, to where the cities of St Louis and Philadelphia now stand.

Climatic zones

During the Quaternary, though the climate as a whole was colder than before, the northern ice sheets at one time advanced, giving glacial conditions, and at another retreated, giving a warmer period (interglacial). These changes have been repeated with varying intensity, at irregular intervals and at least half a dozen times, during the last million years. The presence of land ice in high latitudes led to the establishment of distinct climatic zones (arctic, temperate, and tropical) over the continents. The limits of these zones shifted back and forth with the prevailing state of the ice (advancing or retreating) in what is today (with interglacial conditions) mainly the temperate zone. Thus at one time the climate of a place in southwest France, for instance, might have been subarctic, while in a full interglacial it might have been a little warmer than now.

The glaciations greatly affected the development of life as a whole. But for us this period has a special significance, for it was then that the evolutionary line of primates leading to man made increasingly

39

rapid advances and that man himself appeared on the scene. Man's presence, culture and activities had a considerable influence on the period, so that the Quaternary era is also known as the Age of Man.

The Quaternary is sometimes divided according to archaeological criteria. For instance, the Pleistocene in Europe is often called the Paleolithic, or Old Stone Age, while the Holocene is divided into the Mesolithic (Middle Stone Age) and Neolithic (New Stone Age). Evidence of man's technical and social development (finds of stone implements, dwellings, food refuse, or graves) is as important to archaeologists as fossil finds are to paleontologists. Human paleontologists, indeed, study the rare bony remains of the men themselves. These discoveries show the stage of physique and culture reached by the people concerned—our ancestors—and help us compare Quaternary strata and determine their age.

We must remember that stages of culture, such as that called Paleolithic, do not occur simultaneously all over the world. The Australian aborigines are—or were until recently—*still* at the Paleolithic stage. The advanced pre-Columbian peoples of central and south America were virtually without metals and still in their New Stone Age when conquered by the Spanish in the 16th century. Thus archaeologists should never use human cultural objects as "zone fossils" to date geological strata unless the *local* archaeological sequence is already well known and securely dated.

In some interglacials the climate over most of Europe was very much the same as it is today—damp and fairly warm. Deciduous forests spread widely north and east during these periods, and in suitable undrained places extensive peat bogs formed. Here and in deposits of rivers, sea beaches, lakes, and caves we may find the all-too-scanty relics of Stone Age men and their culture. Sometimes too we come across the bones of the mammals they hunted, as well as the remains of various organic materials—seeds, snail shells, microscopic pollen grains, and so on. All these things can show the kind of natural environment in which these people lived and how they adapted their way of life to it.

During the climatic changes of the Pleistocene both flora and fauna in the northern continents were greatly affected. As the climatic zones shifted slowly southward with the advance of an ice sheet (over 20^0 or more of latitude), so the corresponding successive belts of vegetation also retreated to zones where their constituent plants could survive. These changes occurred over periods of tens of thousands of years; but when the ice eventually retreated the plants reoccupied their former locations. In Europe and western Asia, however, which were subjected to repeated climatic stress, retreat was often blocked by mountains or by the Mediterranean, so that many temperate-zone plants that had first appeared in the Tertiary became extinct.

In Europe, animal species, too, directly or indirectly dependent on certain types of vegetation for survival, either migrated or perished with the plants.

The passage of Atlantic depressions across North Africa, diverted by the glacial high pressure area over central Europe, brought moisture to areas that are now waterless deserts, so that a Mediterranean type of vegetation and fauna developed there.

Vegetation belts

During the glacials three vegetation belts, known as tundra, taiga, and steppe, stretched in front of the Arctic ice sheet for a distance of some 125 to 200 miles. In the tundra, which came right up to the front of the ice sheet, there were mosses and lichens; in particular reindeer moss (actually a lichen). With them occurred the dwarf polar willow, the pygmy birch, and the mountain avens, which is especially characteristic of the ice-age flora in this zone. Azaleas and different species of saxifrage and chickweed also grew there. The taiga belt was dominated by various types of dwarf pines, with a few stunted, small-leaved, deciduous trees like willow and birch, which were accompanied by many types of flowering plants. The dry steppe zone was open country, dominated by grasses and flowering herbs, with trees (chiefly willow, birch, and poplar) only along watercourses and in moister depressions. This belt extended well to the south.

The interglacial flora was of an entirely different character. Some temperate plants died out, but in particular areas some survived—for example, *Brasenia*, a water lily no longer found in Europe; *Dulichium*, one of the sedges; and *Rhododendron ponticum*, a species that now grows wild only in Asia Minor and a few parts of southern Europe. Species common in the interglacials are now rare in the same regions and grow only in certain areas—for instance, the water chestnut, and the yew. In the warmest parts of the ice ages oak, beech, lime, maple, sycamore, ash, alder, hazel, and hawthorn—all broad-leaved deciduous trees—formed a close forest community over most of central Europe.

Mollusks were distributed over wide areas in large numbers. The greatest quantities are found preserved in loess (a wind-deposited dust of glacial origin). Apart from the many land animals with shells —for instance, snails, which still survive—we find many species typical

41

of Arctic and alpine regions. Freshwater bivalves, in particular *Corbicula fluminalis*, at present widespread in Africa, abound in interglacial river deposits.

Elephants and mammoths

The most characteristic members of Pleistocene faunas are mammals, among which elephants are specially prominent. The mammoths *Archidiskodon planifrons* ("flat forehead") and *Archidiskodon meridionalis* ("southern") were found mainly in Asia and southern Europe respectively in the early Pleistocene. The biggest extinct European elephant in the Pleistocene interglacials was *Palaeoloxodon antiquus*, the straight-tusked elephant, which stood 15 to 17 feet high at the shoulder and was a forest-dweller.

The most familiar proboscid was the woolly mammoth, *Mammuthus primigenius*, which flourished in cold environments during the late Pleistocene. It lived in the tundra and was a characteristic member of the glacial fauna. Whole mammoth carcasses in a fairly good state of preservation have been found in permanently frozen ground in many parts of Siberia. They show that *Mammuthus primigenius* had a thick coat of long reddish hair and a close woolly undercoat. This animal has therefore become the best known extinct elephant, and its remains can be found in many museums. In the ice ages the mammoth crossed the Bering Strait into North America, where, during the last of the glacials, it spread very rapidly in the wake of several elephant predecessors. One of these was the related *Mammuthus trogontherii*, which lived from the early Pleistocene to the Last Interglacial and was adapted to a steppe environment. Other relations included *Mammuthus imperator*, the great southern mammoth, and the somewhat smaller *Mammuthus columbi*, which lived in North America during the early Pleistocene.

Interestingly enough, the mastodon, which died out in Europe at the end of the Pliocene, survived all the ice ages in North America. This animal, *Mastodon americanus*, was living on the American continent only a few thousand years ago, at the same time as man. More than 200 mastodon skeletons have been found in New York State alone. In the Pleistocene, contemporaries of the proboscids included the rhinoceroses. One, *Dicerorhinus kirchbergensis* (Merck's rhinoceros) was a companion of the straight-tusked elephant in the interglacial forests of Europe. The woolly rhinoceros, *Coelodonta antiquitatis*, whose remains are even better known, often accompanied the northern mammoth in the colder periods.

42

The genus *Equus*, the horse, is a very important mammal of the Quaternary. It originated in North America, where it had evolved directly from the genus *Pliohippus*, and from there it spread across Asia to Europe. But it died out in North America at the end of the Pleistocene and did not return until brought back there by European colonists. The oldest representatives of the horse in Europe were relatives of the present zebras. They belonged to the species *Equus stenonis*, which lived during the transition from the Pliocene to the Pleistocene. Several races of the wild horse, *Equus caballus*, existed in Europe during the early parts of the Quaternary.

Even the hippopotamus lived in Europe during some warmer parts of the early Quaternary. Its remains, dating from the Last Interglacial, have been found as far north as England, which shows that winters there were free of frost.

One of the most remarkable even-toed ungulates in the early Quaternary was the Giant Red Deer—sometimes wrongly called the Irish Elk—which had spreading antlers rather like those of a fallow deer, with a span of up to 10 feet. It survived in Ireland as late as the early Postglacial. At this time we also find the aurochs, probably the parent of present-day domestic breeds of cattle, and only extinct in the 18th century. Bison roamed the steppes of both Europe and North America, and musk-oxen flourished during glaciations. In the early Quaternary, camels were common in the Mediterranean region. Members of the camel family (Camelidae) of rather different lineage, more like the llama of South America, were also common in the northern part of that continent.

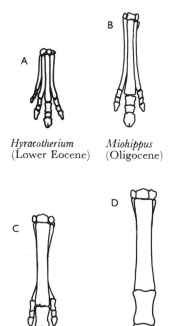

Hyracotherium
(Lower Eocene)

Miohippus
(Oligocene)

Merychippus
(Miocene)

Equus
(Pleistocene and Recent)

24 Development of forelimb of the horse, through the Tertiary to the present, showing the progressive strengthening of one of the central bones and the virtual disappearance of the remaining toes.

Beasts of prey

There were many carnivores in Europe in the Pleistocene, among their most typical representatives being the bears *(Ursus)*. The huge cave bear was common in Europe in the third ice age, but died out during the last. The sabre-toothed tiger *(Machairodus)*, which had dagger-like upper canines, is known from the early Pleistocene and its remains have been found in various parts of Europe. A similar sabre-toothed tiger *(Smilodon)* lived at the same time in America. In the middle and late Pleistocene, Europe was inhabited by the cave

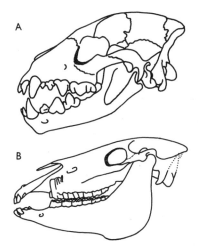

25 Formation of skull and teeth in a carnivore – hyena (A) – compared with those of a herbivore – *Pliohippus* (B), an extinct horse. The hyena uses its large canines for tearing flesh and its cheek teeth for gnawing bones. The horse, a grass-eater, has level teeth, which provide an excellent grinding surface.

43

lion *(Panthera spelaea)*, whose skeleton was almost identical with that of the present-day lion, although it was very different in appearance and much larger in size. It lacked the mane and the tuft on the tail, and its coat was longer and thicker. Prehistoric artists clearly illustrated it in drawings that have been properly studied only during the last few years. Other ice-age beasts of prey included hyenas, foxes, wolves, polecats, and wolverines.

Development of primates

The evolutionary process that started with the primitive forebears of the primates in the Tertiary and ended with the appearance of modern man in the not-so-distant past culminated in the Quaternary. The period of primate development, from its inception in the Paleocene to the evolution of the platyrrhine and catarrhine monkeys, has already been explained. Now let us follow the evolution of man in the late Tertiary and the Quaternary. We know that man came from the Old World catarrhine monkeys, from which a new branch, Hominoidea, sprang, during the later Tertiary era. Anthropologists recognize four main families—Oreopithecidae (not important for the further evolution of man), Pliopithecidae, Pongidae, and Hominidae.

Pliopithecus antiquus, the representative of the extinct family Pliopithecidae, was found in Miocene deposits in the Sansan district in France. It was about the size of a gibbon, but despite certain similarities, it was not the ancestor of modern gibbons, and represents an unimportant sideline.

The Pongidae (apes—unlike monkeys—tailless) have many genera, which are grouped in three subfamilies named after the typical genera —Dryopithecinae, Hylobatinae, and Ponginae. The Dryopithecinae include a great number of extinct genera, mostly represented by fragments of jaws and teeth, but some by more complete skulls and a very few long bones, whose origins go back to the end of the later Tertiary. A typical species of the whole subfamily, *Dryopithecus fontani*, was found in middle Miocene strata near Saint Gaudens in France. Remains of the genus *Gigantopithecus* were discovered in China and India. This genus represents the biggest anthropoid ape known. Its teeth were huge, the diameter of the crown being twice that of the corresponding teeth in man. *Gigantopithecus* lived in steppes and wooded grasslands, and was omnivorous. It finally died out some time during the Pleistocene.

The subfamily Hylobatinae is made up of highly specialized small apes with unusually long arms and long, pointed canines. It includes

44

the gibbons, which live in Far Eastern forests today. The subfamily Ponginae comprises the present-day anthropoid apes, that is, the orangutan, the chimpanzee, and the gorilla, which are evidently descended from Tertiary Dryopithecinae.

The most highly organized and most important primate family is that of the Hominidae, to which man and his most recent ancestors belong. This family took about 14 million years to develop, and the genus *Homo* about three million years. Today the Hominidae are usually divided into four genera—*Ramapithecus* ("Rama's ape"), *Australopithecus* ("southern ape"), *Paranthropus* ("almost a man"), and *Homo* ("man").

Ramapithecus is estimated to have been considerably shorter than modern man (probably less than four feet tall), but unlike the anthropoid apes he already habitually walked on two legs. Remains of skeletons found in Miocene strata in India, China, and Kenya display features entitling him to be included in the direct evolutionary line of man. He is the oldest predecessor of man so far known, and lived in steppes and wooded grasslands about 12 to 14 million years ago.

Australopithecus

The genus *Australopithecus* represents a further stage in man's evolution. The main finds come from Africa. *Australopithecus* was about five feet tall, held himself erect, and walked upright. His skull more closely resembled a modern human skull, and he had a larger brain relatively to his size than the anthropoid apes. He lived in steppes and open spaces and hunted for food. With the bones of this predecessor of man, scientists found very primitive bone and stone implements, suggesting that *Australopithecus* had more than animal intelligence and had started to use his brain and his hands to fashion artefacts. These finds are now estimated to be between one million and three and a half million years old. Many scientists include the controversial species *Australopithecus habilis* ("handy man") in the genus *Australopithecus*, while others believe it is truly *Homo*, because of its tool-making abilities.

The genus *Paranthropus* is closely related to *Australopithecus*, but its members were taller and more robust. It constitutes a blind alley in the evolution of the Hominidae and stopped developing about halfway through the Pleistocene. *Paranthropus* was a forest-dweller and a vegetarian, with huge grinding teeth, and made no implements.

The extinct direct predecessors of modern man and the present human race all belong to the genus *Homo*, whose oldest universally accepted representative is *Homo erectus*.

45

The first trace of *Homo erectus* was found in Java in 1891 by Eugène Dubois, a Dutch physician, who named his find *Pithecanthropus* ("ape-man") *erectus*. In 1936, however, at Modjokerto on the same island, Dr G.H.R. von Koenigswald, a German paleontologist, discovered bones like those of *Pithecanthropus*, but with signs of a lower degree of evolution. They belonged to a bigger, more robust being, and were indubitably older. Von Koenigswald named his find *Homo modjokertensis*. Remains of a similar primitive human type of rather later date had been found in China. At Choukoutien, southwest of Peking, Dr Davidson Black, a Canadian, found one in 1927, whom he named *Sinanthropus pekinensis;* and in 1963, near Lantian, von Koenigswald discovered a jaw of another ancestor of man, older than the Choukoutien remains and belonging to a more primitive being.

In 1935 von Koenigswald had already discovered three huge teeth (of *Gigantopithecus* – see p. 44) in a Chinese drugstore, selling as "dragon's teeth", a highly prized article in traditional Chinese medicine. Thereafter he haunted such places and recognized other teeth identical with those later found at Lantian. Appropriately he named his Lantian find *Sinanthropus* ("Chinese man") *officinalis* ("officina" being the name for the medical storeroom in a medieval monastery).

Further traces of the oldest members of the genus *Homo* were found in Europe. As long ago as 1907 a lower jaw had been discovered near Heidelberg, West Germany, and was named *Homo heidelbergensis*. In 1965 the remains of a being resembling *Sinanthropus* were discovered near Vertesszöllös, not far from Budapest, Hungary.

Similar finds were also made in Africa. In 1954 at Ternifine, near Oran, Algeria, an expedition found the remains of a being with features like those of the Javan *Pithecanthropus*, and named the find *Atlanthropus mauritanicus*. Another similar find was made in the north of Tanzania. In 1953 Dr J.T. Robinson had found remains at Swartkrans, near Sterkfontein, South Africa, which resembled the more primitive Javanese, Chinese, and European finds.

Within the last few years a fossil found at Olduvai Gorge, Tanzania, and known as *Homo habilis* by some authorities, has come to be regarded as an intermediate stage in the development between *Australopithecus* and *Homo erectus*. If this were generally agreed, *Homo habilis* would represent by far the earliest member of the human genus.

Homo habilis, discovered in 1960 by Dr Louis S. B. Leakey and his wife Mary, probably lived about two million years ago. This man-like creature, smaller than a modern pygmy, walked erect, made tools from stone, and lived on a diet of birds, rodents and fish, supplemented by fruit. There is no evidence he had discovered fire, but he may have used stones to build windbreaks.

This was only one of the many discoveries made by the Leakeys, who have been unearthing bones and skulls in East Africa since 1931. In the process, they have revolutionized ideas about man's ancestry by suggesting that the ape and monkey branches must have split off from the Hominidae (p. 45) several million years earlier than was once believed.

Dr Leakey is convinced that Olduvai Gorge was man's original home. Here, in 1959, for instance, he and his wife found parts of the fossilized skull of a primitive kind of man, which they named *Zinjanthropus*, *Zinj* being Arabic for eastern Africa. Potassium-argon dating has established the age of *Zinjanthropus* (often called Nutcracker Man) at one and three quarter million years. And since *Homo habilis* was found in levels below those containing *Zinjanthropus*, Dr Leakey believes these two creatures lived side by side in this part of Africa.

Other outstanding finds made by the Leakeys include: a complete skull of *Proconsul africanus* (p. 169), discovered at Lake Victoria (1948); *Paranthropus boisei*, an Australopithecine skull from Olduvai (1959); *Ramapithecus wickeri*, a late Miocene fossil, perhaps the earliest known hominid (1961); and the remains of *Homo erectus leakeyi*, which were found with some stone hand axes (1971).

Two evolutionary types

After subjecting all the finds from Java, China, Europe, and Africa to a modern comparative analysis, Dr Bernard Campbell, in London, decided that they all belonged to the genus *Homo* and to a single species, which was named *Homo erectus*. Within this species there are two evolutionary types. The more robust, more primitive, and older type includes von Koenigswald's Java man, now correctly entitled *Homo erectus modjokertensis;* the Lantian man, now known as *Homo erectus lantianensis;* the Heidelberg man, now termed *Homo erectus heidelbergensis;* and the African man from Swartkrans, *Homo erectus capensis* ("of the Cape"). The more recent type perhaps includes the other finds, that is, *Pithecanthropus*, now termed *Homo erectus erectus; Sinanthropus*, now *Homo erectus pekinensis;* the Hungarian *Homo erectus*

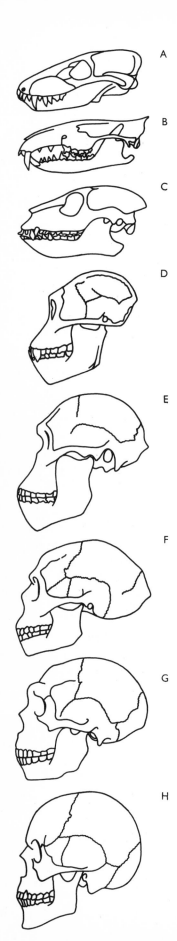

palaeohungaricus; and the other African finds, *Homo erectus mauritanicus* and *Homo erectus leakeyi.*

Homo erectus held himself completely erect, as his straight thighbone shows, and was about five feet six inches tall. He lived in bands, hunted animals, and gathered fruit. He also made different types of stone implements, which he employed for hunting and for cutting up his catch. Evidence that he had discovered how to make and use fire was found at the sites of Choukoutien and Vertesszöllös.

The last link in the evolution of man was *Homo sapiens.* He appeared on the scene quite suddenly, but it is not altogether clear where and how he developed. Traces of his existence have been discovered in many parts of Europe, but his oldest representative is probably *Homo sapiens steinheimensis,* whose remains were found in Germany (1933) and England (1935–36). (The Swanscombe skull, found in Kent, England, was incomplete, but had similar dimensions to that of Steinheim.) Later examples are known from Palestine. Modern man, *Homo sapiens sapiens,* appeared toward the end of the ice ages, some 35,000 years ago.

Until recently Neanderthal man was considered to be a direct predecessor of modern man, but he is now regarded as a distinct race or subspecies of *Homo sapiens,* and has been renamed *Homo sapiens neanderthalensis.* Neanderthalers were about five feet six inches tall and robustly built. They lived from about 200,000 to 40,000 years ago. Modern man could have developed, perhaps in the Near East, from a Neanderthal-like type of ancestor, during the Pleistocene. His further development can be followed from traces of his increasingly elaborate culture, which culminated in our modern civilization.

26   Evolution of the skull from reptile to man (not to scale):

A *Thrinaxodon,* a mammal-like reptile

B *Zalambdalestes,* a primitive insectivore

C A lemur, *Notharctus*

D *Rhinopithecus,* an Old World monkey

E *Australopithecus,* an early Hominid

F *Homo erectus erectus,* the Java man

G *Homo sapiens neanderthalensis,* Neanderthal man

H *Homo sapiens sapiens,* modern man

48

Precambrian era

(subdivided into Archaeozoic and Proterozoic periods)

Primitive arthropods,
coelenterates,
worms, and sponges

2600 million
years ago

Primitive plants
Primitive cellular organisms
Primitive micro-fossils
First primitive micro-organisms
Biological evolution
Origin of life
Formation of proteins
Development of organic
    compounds
Development of atmosphere
    and hydrosphere
Formation of earth's crust
Formation and development
    of earth

4600 million
years ago

Formation of solar system

Development of Universe
    and galaxies

The earth's surface was formed some 4600 million years ago. At first, it must have been quite bare, with an average temperature of about 0°C. Since there were no clouds, day and night temperatures would have differed widely, perhaps by 50°C. In its earliest solid form our planet was one vast inhospitable desert and looked very different from what it does today. Peaks and hills were of sharper outline;

slopes were more abrupt. Without atmosphere or water, there was nothing to weather the rocks and wash the debris down into the valleys; and without water-borne sediment there was no soil. But in places the earth displayed very high volcanic activity. As the crust was constantly broken down and remade, whole regions rose and sank. New rocks appeared, only to be remelted, so that their entire structure altered.

Some 4000 million years ago dense clouds hung over the bare earth, often blocking out the sun's light. The earth's surface groaned under gigantic storms, while streams of molten lava gushed forth in many places. Volcanic activity released increasing amounts of gas and vapour from inside the earth, and these gradually formed an atmosphere too heavy to escape into space. Water began to appear in depressions

in the earth's surface, first as pools, then as large sheets of water. In time these united to form the first oceans, while heat and cold, wind and water, disintegrated the rocks and produced the first sediments. Life itself did not yet exist, but the organic compounds needed to build living matter were already present.

Life originated on earth about 3500 million years ago, although the oldest evidence we have for its existence does not go back more than about 3200 million years. In time, the first plants appeared, followed by the first invertebrates (animals without backbones). By about 2000 million years ago the shallower seabed was covered with different primitive plants. These included blue-green algae, simple plants that marked an enormous step forward in the evolution of life, since they contain a complex green pigment called chlorophyll. This enables plants, using the energy of sunlight, to manufacture substances from non-living matter, such as carbon dioxide and water, for their food and growth. In the process, known as photosynthesis (putting together by means of sunlight), green plants release oxygen, probably not present in the atmosphere before their time.

Paleozoic era

		Trilobites and primitive corals disappear
		Development of "mammalian" reptiles
PERMIAN PERIOD	280 million years ago	
		Appearance of first gymnosperms Reptiles appear First winged insects Development of amphibians First forests
CARBONIFEROUS PERIOD	345 million years ago	
		First amphibians Varied development of fishes First insects First tree-like plants More advanced plants
DEVONIAN PERIOD	395 million years ago	
		First jawed fishes First land animals (arthropods) Land plants appear Many coral reefs
SILURIAN PERIOD	445 million years ago	
		Development of cephalopods Development of arthropods Development of coelenterates Development of graptolites Jawless fishes
ORDOVICIAN PERIOD	500 million years ago	
		Development of echinoderms Development of trilobites Development of invertebrates Traces of vascular plants Development of algae
CAMBRIAN PERIOD	570 million years ago	

Early Cambrian seas swarmed with life. The trilobites shown here—*Paradoxides gracilis* (foreground) and *Ellipsocephalus hoffi* (smaller specimens, right)—were among the most common animals of the period. So-called because they were divided lengthwise into three lobes, these flat-bodied crustaceans crawled about the seabed

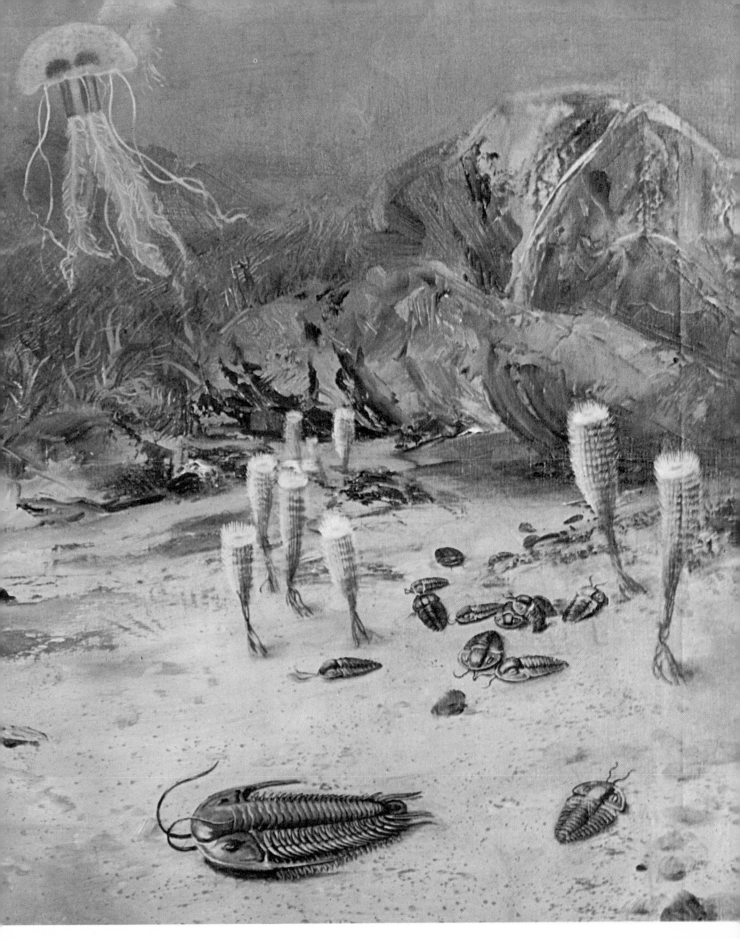

in search of food. Most trilobites were about an inch long, but others were so small that their fossil remains can be seen only with a microscope. The picture shows also jellyfish, and various forms of sponge anchored by their stalks to the seabed.

Seabed dwellers of Cambrian times depicted here include three species of trilobite—
Paradoxides gracilis (foreground), *Conocoryphe sulzeri* (middle, larger specimen), and
Ptychoparia striata (left, two specimens). Sea lilies, or crinoids, which channeled
food along their waving arms to their mouths, are anchored to the bottom by their
long stalks (right). A group of brachiopods, marine animals that have two valves,
occupy the foreground, right.

By the Ordovician period, when shallow seas covered large areas of the world, primitive plants such as *Boiophyton pragense* had begun to adapt themselves to life on the relatively dry land. As the illustration shows, *Boiophyton* had already developed a stiff stem to hold it upright against gravity; equally important, it was vascular—it had developed a system of vessels or tubes to carry water through its body.

Ordovician trilobites included the striking species *Cryptolithus ornatus*, shown here. Its head, which widened toward the tip, had a semicircular shield with a flat, closely perforated border. Echinoderms ("spiny-skinned") were characteristic inhabitants of the sea. To the left is a primitive species that carried its bulging body on a short stalk. Another species (right) was also rounded or pear-shaped, but anchored itself by the tapering end of its body. The sea lilies (extreme left) belong to the same group. To their right lies a conical shell *(Elegantulites elegans)*. Brachiopods (white shells, right) furnish the most abundant invertebrate remains in Ordovician sediments, however. They lived only in the sea and their bodies were usually enclosed in a limy bivalve shell. Cephalopods (early relations of the cuttlefish) already swam in the water. Jellyfish and small passively drifting animals were numerous. Parts of the seabed were overgrown with algae, which provided food for the Ordovician fauna.

Living things abounded in the warm coastal zone of Silurian seas. The seabed was covered with flat corals of the genus *Favosites* (bottom right and middle), and with skeletons of large stromatopores ("spreading sponges")—bottom left—some of which, when alive, formed high, bulging tufts (background). Spherical algae of the genus *Ischadites* (left) and different species of wrinkled corals flourished in these waters. Various gastropods searched for food among the corals and dead sponges. Cephalopods, in tube-like shells, whose arms appear to sprout from their heads, roamed the mid-water for prey.

Nautiloids—cephalopods whose shells are built like those of the present-day nautilus—were typical dwellers of the late Silurian seas. Some species reached a length of 12 feet. Their shells came in a variety of shapes and sizes. Some were long, some spiral; some had curved or bent shells, while others were straight or slightly twisted. Also in the picture is a group of sea lilies (right); in front of them lie rounded clumps of flat corals and a few skeletons of goblet-shaped wrinkled corals.

Eurypterids, fast-swimming sea scorpions that flourished during the Silurian period, were probably carnivorous and may have preyed on the first vertebrates. The two main genera were *Pterygotus* and *Eurypterus*. *Pterygotus* reached a length of well over six feet and had huge pincers for its first pair of limbs; *Eurypterus* was less than a foot long and its sixth pair of limbs had evolved into swimming organs.

At the end of the Silurian period, swamps and marshes beside the sea were already occupied by low vegetation composed of the most primitive types of vascular cryptogams *(Psilophytales)*, plants that reproduce, like ferns, by wind-dispersed spores. Some genera were leafless, but others already bore leaves. On the right, in the water, is the low, minute *Zosterophyllum* (a form between ferns and club mosses); to the left of that, on the land, grows the tall *Psilophyton*, with the low, creeping, spidery *Sciadophyton* below it. On the left, by the water, is *Protolepidodendron*, the forerunner of the present club mosses, and in the background are clumps of the creeping *Drepanophycus*, which looked rather like a club moss.

Horsetails, club mosses, and ferns were the most characteristic plants of middle Devonian times, although primitive plants such as *Psilophytales* (see previous page) still survived. The various kinds of club moss included *Protolepidodendron*, a tall plant with bent branches (left, in middle and far distance) and the stumpy *Barrandeina* with its three branches (foreground).

Above: Small members of the genus *Pteraspis*, a jawless type of armoured fish with a long snout, scatter as *Osteolepis*, a bony freshwater fish equipped with primitive air sacs (top right), moves in to attack. *Osteolepis*, which lived in the Devonian period, some 350 million years ago, headed the long evolutionary line that eventually led to man. On the left is a pair of ostracoderms—the flat-bodied, armoured *Drepan-aspis* in the foreground, with *Psammolepis* just behind it.

Opposite: Three primitive, heavily armoured *Placodermi*, Devonian contemporaries of the ostracoderms. Their heads and bodies were encased in thick armour. *Antiarchi*, the species shown here, lived near the bottom and had paddle-like limbs, with which they "rowed" themselves along.

Some sea-going placoderms, such as *Duncleosteus*, reached huge dimensions in the course of evolution. The specimen shown here pursuing a small shark measured almost 40 feet in length. It had well-developed jaws formed of paired plates of bone, with a large, fang-like projection in front and sharp edges at the back. The front of the trunk was heavily armoured, and the hind part was either bare or covered with small bosses or scales.

Acanthodii, the oldest primitive bony fishes, lived in fresh water in most parts of the world, and are known from Silurian and Permian strata. Generally, they were about four inches long, but types that migrated to the sea were much larger. Their bodies were armour-plated.

Bones of various stegocephalians, the first dry-land quadrupeds, have been found in upper Devonian strata in the eastern part of Greenland. The picture shows *Ichthyostega*, an amphibian that grew to a length of about three feet and had a fish-like body and well-developed limbs. The flora of that time was very varied and included the tall-trunked club moss *(Cyclostigma)*, *Pseudobornia* (the horsetails in front of the nearer amphibian), *Sphaenophyllum* (the creeper on the stump), and *Archaeopteris* (the ferns near the water).

Plants flourished in swamps and lagoons in the warm, damp Carboniferous. In
the background, right, is a giant club moss—*Lepidodendron* ("scale-tree"); on the left
stands a tall, plumed *Sigillaria*, with long, thin leaves. The two tallest trees pictured
here are cordaits. The fan-leaved *Psaronius* (middle) is a typical representative fern.

During the Carboniferous period, Hercynian folding gave rise to extensive basins that were soon overgrown with impenetrable forests. Today there are coalfields

in these places. The "trees" of Carboniferous forests consisted mainly of huge club mosses, calamite horsetails, cordaits, and tree ferns.

Left: Carboniferous forests beside shallow lakes swarmed with life. Living and rotting vegetation provided food for insects, spiders, crustaceans, and many other invertebrates, which in turn were preyed on by the first amphibian vertebrates, the stegocephalians (lower picture). The illustration here shows a giant dragonfly, *Meganeura*, wingspan 28 inches, which lived on small insects.

Below: A group of stegocephalians. *Branchiosaurus*, small and dark, is resting on a rock (right), while below it lies *Microbrachis*, with a long body and short limbs; above it two specimens of *Urocordylus* are hunting insects. *Dolichosoma* is swimming in the water (left).

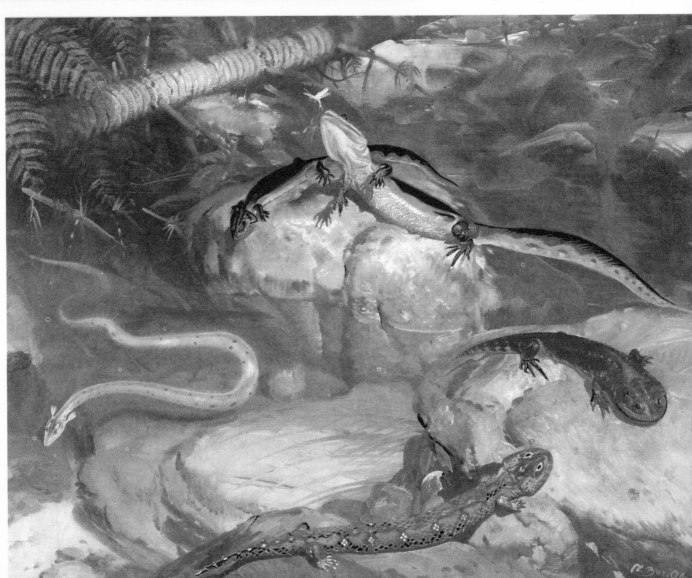

Left: *Diplovertebron*, only 24 inches long, was an amphibian. Although it had well-developed limbs equipped with five digits, it sought its food mainly in the water. It forms an important link in evolution, because its body structure resembled that of a reptile.

Below: The freshwater shark *Pleuracanthus*, which measured up to 28 inches in length, pursuing members of a group of primitive fishes that lived in Carboniferous lakes and rivers. *Pleuracanthus*, whose skeleton was made mostly of cartilage, had a long conspicuous spine behind its head.

Uprooted by tornadoes, the trees of the Primary era fell into the swamps and in course of time were carbonized. The picture shows a *Sigillaria* being snapped, while

lepidodendrons and cordaits bow to the force of a gale. Whole fans ripped from tree ferns sail through the air.

Edaphosaurus (below) is a relatively well-known representative of the "mammalian" reptiles, an independent group from which the first mammals evolved at the end of the Triassic period. It lived in the late Carboniferous and early Permian periods, a time when climatic changes were having a marked effect on the fauna, especially on vertebrates. As spells of drought grew longer, some stegocephalians slowly adapted themselves to the altered conditions in dry regions. Instead of laying their eggs in water, they hid them in rotting vegetation or sun-warmed sand. The first species of edaphosaurs were small, but by the Permian period their successors measured over 10 feet in length.

During the Permian period, at the end of the Paleozoic era, great changes occurred in the earth's climate and appearance. The warm zone spread in the northern hemisphere, where extensive dry deserts appeared. The rocks formed at that time were stained red by iron oxides, the result of intense heating by the sun of a surface devoid of vegetation cover. The old types of plants and animals died out. Ferns were no longer the dominant plant species; they were succeeded by gymnosperms, including conifers such as *Walchia*. These were very widespread and, unlike ferns, were able to grow in dry areas. Reptiles grew to dominance among vertebrates, because their special adaptations enabled them, also, to flourish in the drier climate.

79

The wetter intervals of the Permian period, a time of extremely violent change, were marked by the growth of luxuriant vegetation, mainly of types already present in the Carboniferous, which flourished round the rivers and lakes.

Above: *Seymouria* (bottom right), whose body structure displays features typical of both amphibians and primitive reptiles, was an amphibian, but is often regarded as the possible ancestor of the reptiles. About two feet long, it had sharp teeth and probably lived on any smaller animals it could catch on land or in the water. (The first fossil remains of this animal were found in 1901 near Seymour, Texas.) *Diadectes* (top), an herbivorous reptile, measured six feet in length. It was representative of the important evolutionary line that probably evolved from *Seymouria*-like ancestors and led, via different groups of reptiles, to the birds.

Left: A group of amphibians that lived permanently in water in the early Permian swamps and lakes of Europe and Asia. Their body structure displayed marked reptilian features, however, showing that they were close to the line that divides the two groups. *Discosauriscus* (right and top left) measured about 18 inches and had a longer, more tapering head than its smaller cousin *Letoverpeton* (bottom left), which was almost half the length and had a wide, short head. Above: Discosauriscids were evidently the prey of *Chelyderpeton*, a large, lizard-like amphibian that reached a length of four feet.

Above: *Moschops*, an herbivorous therapsid that lived by South African rivers in the middle of the Permian period. More than six feet in length, it had a third eye, sensitive to light, on top of its large, wide skull.

Left: *Dimetrodon* (top) and *Varanosaurus* (bottom), representatives of two of the three groups of Pelycosauria, "mammalian" reptiles from which the evolutionary line to the first mammals can be traced. (A representative of the third group, *Edaphosaurus*, is illustrated on p. 78.) *Dimetrodon*, a powerful carnivore, supported a huge dorsal fan, which may have acted as a heat receptor.

Scutosaurus, which was about the size of a fully grown bull, represents another family of primitive cotylosaurian reptiles that lived in Permian times. Its heavily armoured, stocky body had a relatively small head, a short tail, and stumpy, splayed legs.

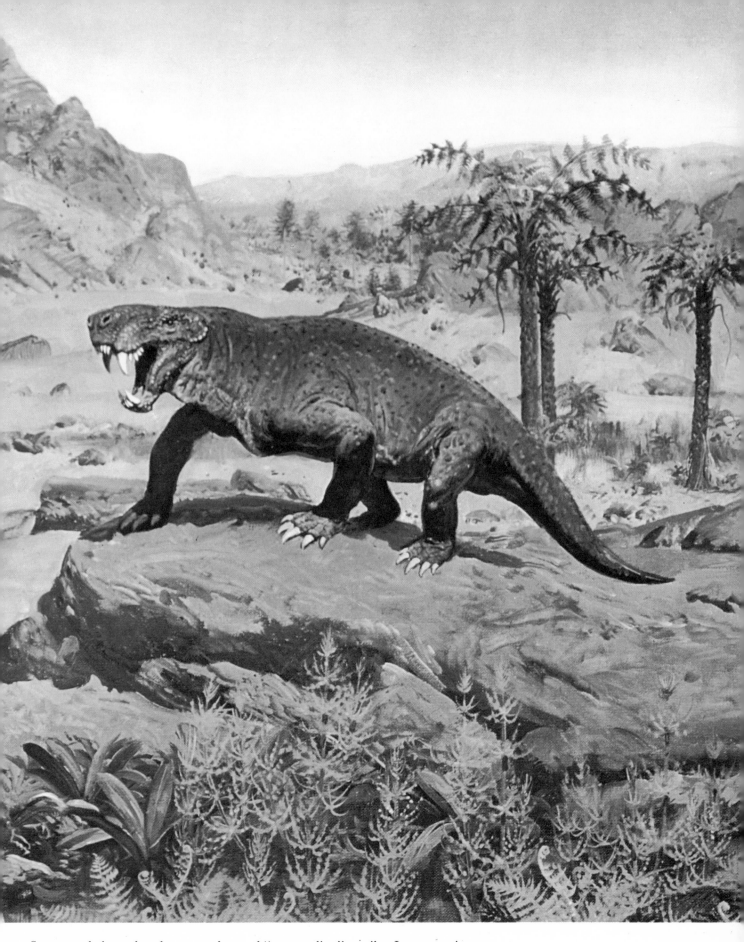

Sauroctonus belonged to the more advanced "mammalian" reptiles. It was carnivorous and had powerful teeth. Its body was shaped rather like that of *Dimetrodon* (p. 84).

Bones of this freshwater reptile, *Mesosaurus*, which measured about three feet in length and lived on fish, were recently found in Antarctica (once part of the southern continent of Gondwanaland). The animal had a very long skull, large numbers of teeth, and a flat-sided tail.

Mesozoic era

Dinosaurs extinct

Primitive fishes disappear

Disappearance of ammonites

Development
of flowering plants

Development
of primitive mammals

Development
of giant horned saurians

Development of birds

CRETACEOUS
PERIOD

136 million
years ago

Development of winged reptiles

Primitive birds

Tailed amphibians

Dinosaurs become dominant

JURASSIC
PERIOD

195 million
years ago

First mammals
First frogs
Development of reptiles
Development of two-gilled
cephalopods
Development of ammonites
Development of modern corals
Development of gymnosperms

TRIASSIC
PERIOD

225 million
years ago

By the outset of the Mesozoic era, cycads (left foreground) had become the most important of the gymnosperm plants. They were either stumpy, with round or

barrel-like stems, or had tall, slender trunks. All of them were topped by a tuft of palm-like leaves. Conifers were also becoming widely distributed.

Landscape typical of the early Triassic period in Europe, a time when the climate was so dry that vegetation grew only in places where there was moisture. Beside the lake stands a six-foot-high *Pleuromeia;* its slender trunk, marked by leaf scars, was crowned with short, tough leaves and ended in a conspicuous cone. The sand beside the lake is marked with the tracks of the reptile *Chirotherium,* an animal that is totally unknown except for its footprints.

Above: *Mastodonsaurus*, a huge amphibian stegocephalian, found in Triassic strata in Europe and North Africa, had a flat skull four feet long. The biggest amphibian of all time, it was mainly aquatic and lived on fish. It died out at the end of the Triassic.

Left: *Triadobatrachus*, one of the first primitive frogs, which appeared in the lower Triassic. It measured about four inches overall and was probably an offshoot of the line that led to modern types of frog.

93

Mixosaurus (above) is one of the best known of the primitive types of ichthyosaurs ("fish-lizards"). These animals, which were better adapted to an aquatic existence than any other reptile, appeared in the Triassic, reached their heyday in the Jurassic, and disappeared at the end of the Mesozoic era. *Mixosaurus* had a shorter snout than Jurassic ichthyosaurs (cf. p. 104), and its long jaws were armed with numerous teeth. It reached a length of over six feet.

Nothosaurus was a marine reptile very well adapted to life in the sea. A contemporary of the first ichthyosaurs and placodonts (p. 97), it is regarded as the ancestor of the plesiosaurs (p. 110). It measured about 10 feet in length and its mode of life was probably very similar to that of the seals today, its main food being fish. Its remains have been found in Triassic sediments in Europe, southeast Asia, and North America.

The first representatives of the testudinates (tortoises and turtles) appeared in the middle of the Triassic. *Proganochelys*, one of the oldest, was about 28 inches in length. Its remains were found in West Germany. Testudinates have undergone few changes. They survived all the natural catastrophes that destroyed most of the other reptile groups and are almost as abundant today as in the past.

Below: Placodonts, early Triassic marine reptiles that lived mainly on mollusks. Their abdomens were protected by a bony plate and they had a functional third eye on top of the head. *Placodus* ("plate-toothed"), seen here, had a rounded snout and large numbers of teeth, whereas *Placochelys* (not illustrated) had a tapering snout, fewer teeth, and thicker armour. The differences between their respective snouts are illustrated opposite, *Placodus* appearing above *Placochelys*.

Left: *Saltoposuchus*, a thecodont ("socketed-teeth") ancestor of the saurischian ("lizard-hipped") dinosaurs, which developed during the Triassic. Small reptiles with sharp teeth and lizard-like bodies, thecodonts walked on their hind legs, using their long, thick tails to keep their balance. *Saltoposuchus* was about four feet long and is known from European upper Triassic rocks. Note the disproportion between its short forelimbs and long, powerful hindlimbs, a disparity that increased during the course of evolution. The thecodonts are extremely important, as they were the stock from which the crocodiles and birds, as well as the dinosaurs and winged reptiles, developed.

Below: The toothless *Henodus*, one of the most remarkable of the placodonts. This curious reptile, which measured well over three feet, was found in Triassic strata near Tübingen, Germany. It had a square head and a short body encased in a turtle-like shell.

The best known forerunner of the crocodiles is *Protosuchus* (above), one of the the-codonts, which measured about three feet in length. Its body was covered with armour-plating reinforced down the middle of the back by a double row of bony plates. Judging by the structure of its limbs, *Protosuchus* was probably both a good runner and a good swimmer. Its skeleton displays many features typical of the true crocodiles, and its sharp, pointed teeth show that it was a flesh eater. It lived in North America and Europe in the late Triassic and early Jurassic periods.

Dicynodonts ("with two dog-teeth") were among the more advanced forms of "mammalian" reptile during the Permian and early Triassic period. *Lystrosaurus* (depicted below) measured about four feet in length and was probably aquatic, although it was also able to walk on dry land. Most of its teeth were modified to a kind of beak like a turtle's.

Pterosaurs (winged reptiles) were the first vertebrates to conquer the air. Like *Rhamphorhynchus* ("beak-snouted"), shown here, they evolved from thecodont forebears toward the end of the Triassic; but their development culminated in the Jurassic. They died out at the end of the Cretaceous. *Rhamphorhynchus* had a wingspan of up to six feet, a long skull and neck, a short body (about 20 inches long), and a long, thick tail ending in a diamond-shaped flap of skin. The fourth digit of its forelimbs was extremely long and acted as a support for the flying membrane, which was attached to its body. Its hindlimbs were stunted. Pterosaurs had hollow bones filled with air, like birds, and their bodies were thinly covered with hair.

Jurassic landscape scene. This period, which started some 195 million years ago, was wetter and warmer than the Triassic. The many swamps and freshwater lakes

were inhabited by the first frogs; the luxuriant vegetation consisted mainly of gymnosperms, with conifers and cycads (palm-like trees) predominating.

Two streamlined ichthyosaurs from the Jurassic period. Members of this genus *(Stenopterygius)* had small heads and relatively narrow fins. Embryo skeletons, as well as food remains, have been found in the body cavities of several specimens. Some were quite large: for instance, a fully developed embryo 20 inches long was found in the body of a female measuring about 10 feet.

Eurhinosaurus longirostris, another type of ichthyosaur, is notable for its short lower and extremely long upper jaw. No satisfactory explanation has so far been found for this development. Although ichthyosaurs were very numerous and widely distributed, they did not, for some reason, survive the end of the Mesozoic era.

Pterodactyls, winged reptiles that appeared in the late Jurassic period, lived in flocks and ate fish and insects. They had short tails, wide wings, tapering skulls, and a few teeth in their beak-like jaws. They frequented the seashore and may have slept hanging from the rocks or branches, rather like bats.

Dimorphodon ("two-form-toothed") measured just over four feet, including its tail, and had a short, lightly built body and strong limbs. Most striking was its bulky yet light skull, about nine inches long. It had a wingspan of some five feet and must have been a carnivore. Pterosaurs died out during the Cretaceous, probably because they could not compete with the birds, which had a superior anatomical structure and perhaps greater flying ability. No one knows just how pterosaurs flew; perhaps they merely soared or glided.

Right: The first true crocodiles appeared in the early Jurassic period, on sandy seashores. *Steneosaurus bollensis*, shown here, was one of the commonest and grew to between 13 and 20 feet. Its five-toed forelimbs were much smaller than its four-toed hindlimbs. Both pairs of feet were webbed and the first three toes had claws. *Steneosaurus* abounded in the Jurassic seas of Europe, North Africa, and America.

Below: *Metriorhynchus brachyrhynchus* was explicitly adapted to life in the sea. It measured nine feet in length and had paddle-like limbs. It swam mainly by means of its tail.

These two pages illustrate the two distinct types of plesiosaur—the short-necked *Peloneustes* (above) and the long-necked *Plesiosaurus* (opposite). These marine reptiles occurred throughout the Mesozoic. Both types were about 10 feet long.

Proavis, the animal depicted here in imagined reconstruction, represents an attempt by paleontologists to suggest what the link between reptiles and primitive birds may have looked like. They presume that such a creature still had separate "fingers", but that the reptilian scales had already been modified as feathers. The reptilian head (see close-up) would still be covered with coarse scales and well armed with sharp teeth.

Archaeopteryx (right), the earliest primitive bird, was about as big as a crow. But mouth, wings, and tail showed that it was closely related to reptiles. The jaws were armed with small teeth, and were covered with scales (see detail, below). The wings had three clawed toes, which were used for climbing, and the long tail took up about half the spine. *Archaeopteryx* ("ancient wing") could certainly fly, but the relatively flat breastbone showed that its wing muscles must have been too weak for really sustained flight. Being both warm-blooded and feather-covered, it had a great advantage over the reptiles, for it could be more active and withstand changing temperatures better. It probably ate berries, insects, and worms.

Jurassic landscape scene. The shore of the lagoon is thickly overgrown with ferns and other plants. A small dinosaur, *Compsognathus longipes*, and two specimens of *Archaeopteryx* (p. 113) are searching for food. *Compsognathus* was about the size of a chicken, and perhaps lived on insects and berries. Equipped with powerful hindlimbs, it could run rapidly over the grassy steppes and sandy shores.

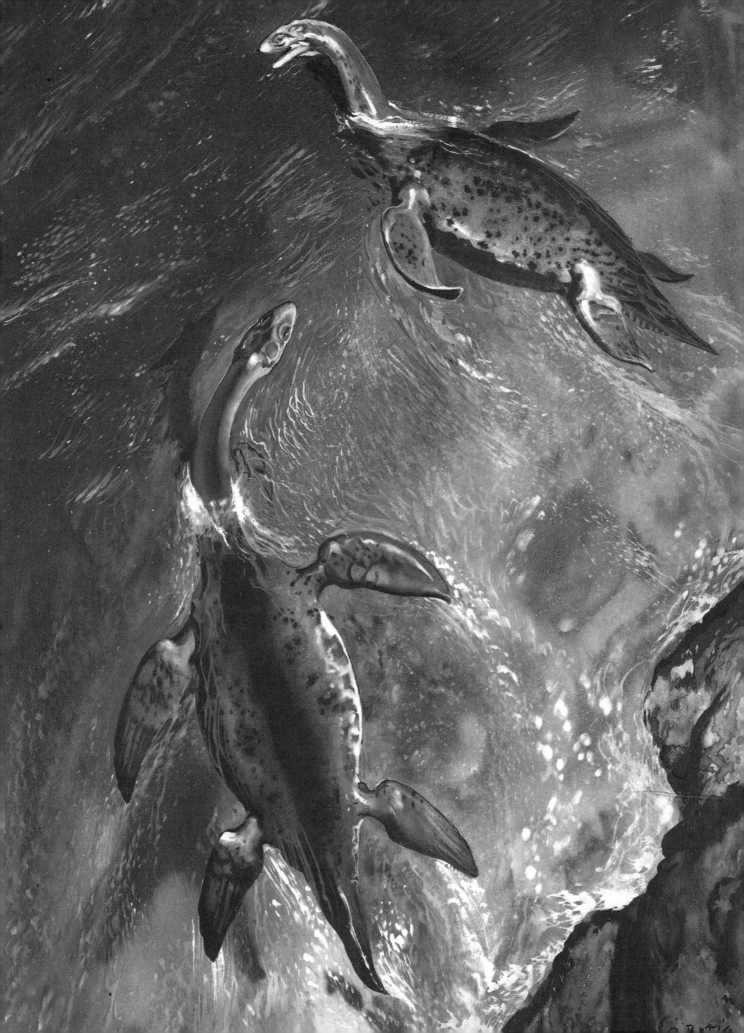

Above: Stegosaurs, 20-foot-long, armour-plated dinosaurs that lived during late Jurassic times. Down their backs ran a double row of thick, bony plates, perhaps to protect the spinal column. The short, thick tail, which the animal used for defence, was armed with two pairs of long, bony spikes. *Stegosaurus*, which was a vegetarian, had an extremely small head and a brain about the size of a walnut. In addition, it had two other, larger, "brains"—one located in its lumbar region, the other between its shoulders. Left: *Cryptocleidus oxoniensis*, long-necked plesiosaurs, which were common in Jurassic seas and no doubt devoured any creature small enough to be swallowed. (See also p. 110.)

Brontosaurus ("thunder-lizard"), a member of the dinosaur family, was the biggest land animal that has ever existed. It weighed about 20 tons and reached a length of 60 feet. Paleontologists at one time believed that a creature of such tremendous bulk must have spent most of its life in lakes and swamps, where it could use water to support its body. They argued that on land the ends of its leg bones would have been crushed by the enormous weight. But more recent evidence (from traces of

footprints) suggests that big dinosaurs could move in water too shallow to reach their bellies, and on muddy ground. In relative terms, *Brontosaurus* had the smallest brain of all the vertebrates. The main function of the brain, which weighed only a pound, was to work the jaws and warn the animal when food or danger were near. In addition, the big dinosaurs had a swelling at the base of the spinal cord, which acted as a second brain, and worked the muscles of the hind legs and tail.

Diplodocus carnegii, at 90 feet the largest of all the dinosaurs, was not as heavily built as *Brontosaurus*. It had a slender neck shaped like a swan's and a long tail with a whip-like tip. *Diplodocus* lived in late Jurassic swamps, about 150 million years ago, in the

area now covered by Wyoming, Utah, and Colorado. The species was named in honour of Andrew Carnegie, the American industrialist and philanthropist, who in 1909 financed the excavation of one of them.

Brachiosaurus brancai, pictured on the two preceding pages, was a giant herbivorous saurian (75 feet long, 40 feet high) that lived in Africa, North America, Europe, and eastern Asia during the Jurassic and early Cretaceous periods. Four thick limbs supported its huge body. As the forelimbs were longer than the hindlimbs (the name means "arm-lizard"), it differed significantly in its proportions and attitude from both *Brontosaurus* and *Diplodocus*. Its eyes were set high in its head, with the nostrils raised on a crown. When danger threatened, it could still breathe while remaining almost completely submerged.

Triconodon, a primitive mammal known only from its fossil teeth and jaws, evolved from advanced "mammalian" reptiles at the end of the Triassic. Up to two feet in length, it was probably a carnivore and lived on small reptiles like *Sapheosaurus*, the prey shown here. A warm-blooded animal, it had a hairy coat that must have helped to protect it from extremes of heat and cold. Although *Triconodon* may have laid eggs like its reptilian ancestors did, it probably suckled its young.

The Jurassic flora was tropical in character. Gymnosperms, mostly cycads, were the chief type of plant. The palm-like *Williamsonia*, which reached a considerable height, belonged to the gymnosperm group of the Bennettitales, whose first representatives appeared in the Triassic. Conifers and ginkgoes (of the same genus as the modern maidenhair tree, its sole survivor) were also widespread. Part of the crown of a ginkgo, showing the deeply indented leaves, can be seen on the left, at the edge of the picture.

The forests of Jurassic times were filled with gymnosperms. Predominant among them were the tall, pointed ancestors of the sequoia, and a giant conifer that had a thick trunk and luxuriant crown.

The large dinosaur *Iguanodon* ("iguana-tooth") stood over 15 feet high and was over 30 feet in length. These reptiles, which weighed three or four tons, walked on their powerful hindlimbs and browsed on trees and bushes. Their flat, notched teeth were constantly being worn down and replaced. Their dagger-like thumbs were used only for defence. Although *Iguanodon* was first discovered in Sussex, England, in 1822, the most spectacular find was made near Bernissart, Belgium. Here 23 almost complete skeletons were unearthed in a coal mine in 1877.

Pteranodon ("wings, no-teeth"), the largest flying reptile of all time, had a wingspan of over 25 feet, a long, sharp bill, and a bony crest on its head. As its legs are supposed to have been too weak to support its body, it possibly fed its young from the air. The bill was perfectly adapted for catching fish and other marine animals. The wings were attached to the forelimbs, the "little finger" forming the leading edge beyond the hand, and were stretched out by the hindlimbs.

Tylosaurus dyspelor, a 35-foot-long marine reptile belonging to the Mosasauria group, threatens *Elasmosaurus platyurus* (right), a plesiosaur some 44 feet in length. These reptiles lived in the upper Cretaceous seas of North America. Tylosaurs had skulls more than three feet long and used their tails as the main swimming organ.

Left: The giant turtle *Archelon ischyros*, which was about 14 feet long and must have weighed two or three tons, lived together with other marine reptiles such as mosasaurs and plesiosaurs in the upper Cretaceous seas of North America. Adaptation to an aquatic environment led to considerable reduction of the shell. Its powerful limbs ended in extremely long "fingers" joined by webs.

One of the "duck-billed" dinosaurs, *Corythosaurus casuarius* (above), reached a length of up to 30 feet. The members of the genus shown here, which were swamp dwellers, had a striking hollow crest like a helmet on their heads. Studies of the pelvis (in most vertebrates the part giving the best evidence as to sex) of these reptiles suggest that only the males possessed this distinction.

Opposite: *Monoclonius* ("single shoot"), like *Protoceratops*, one of the "horned" dinosaurs, had a bony collar that bordered the back of its skull. It had only one horn over its nose (hence the name) and was about 17 feet in length.

Protoceratops andrewsi (left) was a herbivorous dinosaur whose life history is known from egg to adulthood. Between six and seven feet long, it had a large skull, which ended in a hooked beak, and a rounded "collar" above the neck. Nests of fossilized eggs of *Protoceratops* have been found in the Gobi desert, Mongolia.

Of all the horned dinosaurs, *Styracosaurus* ("spiky lizard") was the most fearsome in appearance, though probably harmless. Its nose was armed with a huge horn almost two feet long, and six thick, pointed horns were arranged in a semicircle around its perforated bony collar. These horns must have given it efficient protection against enemies. The head was seven feet long and five feet wide. *Styracosaurus* lived at the same time as *Monoclonius* (previous page), which it resembled in the basic structure of its skull. *Styracosaurus* was extinct by the end of the Cretaceous.

The biggest and most frequently found horned dinosaur is *Triceratops* ("three-horn-face"). It had three great horns, one on its nose and two above the eyes; the latter resembled a bison's horns. *Triceratops* was about 20 feet long and 8 feet high. Its skull was about 7 feet in length, but much of this was accounted for by the protective bony collar. Many species lived in the region now covered by Wyoming, Montana, and Colorado, where there were extensive swamps and plenty of vegetation. Like the other dinosaurs, *Triceratops* died out at the end of the Cretaceous. The illustration shows *Triceratops prorsus*.

Like its contemporary, *Tyrannosaurus rex*, *Gorgosaurus* (below) was a huge carnivorous dinosaur. Some 27 feet long, it probably attacked smaller saurians of the genus *Scolosaurus*, which were protected above by thick bony plates and large spikes. The spiked tail, too, would have been a useful weapon to ward off predators.

Tyrannosaurus rex about to attack a "duck-billed" *Trachodon*. Together with *Gorgosaurus* (opposite), *Tyrannosaurus* ("tyrant-lizard") was the biggest biped dinosaur that ever existed. Its bones were found in upper Cretaceous sediments in Montana in the United States. About 40 feet long and 17 feet high, it was carnivorous, and its great jaws and dagger-like teeth were well equipped for tearing flesh. Footprints as well as skeletons have been found. One of them was some 28 inches long and over 31 inches wide; the distance between strides was 12 feet 6 inches. The appearance of *Tyrannosaurus* must have struck terror into the hearts of other dinosaurs. Although of almost the same length, *Trachodon* would have been virtually helpless against such an enemy: its only hope of safety would have been to escape into water, where it could swim. *Trachodon* had a long skull and a flat snout. It had about 1000 teeth, arranged in rows. In the background of the picture, dinosaurs of the genus *Ornithomimus*, looking like plucked ostriches, are being pursued by another tyrannosaur.

Primitive birds with teeth, like *Ichthyornis* (right) and *Hesperornis* (below), lived on the shores of North American upper Cretaceous seas about 60 to 70 million years ago. *Hesperornis* ("western bird") was over three feet long; it did not fly but was an excellent diver and swimmer. *Ichthyornis* ("fish-bird") was about the size of a pigeon, had well-developed wings, and was a skilled flier. Like *Hesperornis*, it lived on fish. Both animals were already true birds, but they still had teeth, testifying to their reptilian origins. By the end of the Cretaceous they were both extinct.

Mosasaurs, enormous marine reptiles, abounded in late Cretaceous seas all over the world. These carnivorous lizards were wonderfully adapted to an aquatic mode of life, and many of them measured up to 40 feet in length.

The Asian dinosaur *Tarbosaurus bataar* was closely related to the late Cretaceous carnivorous dinosaurs of North America. In appearance and size it resembled the tyrannosaurs. It walked on powerful hindlimbs and used its tail to balance itself; the forelimbs were stunted and had only two "fingers". Its jaws were armed with large pointed teeth. In recent years, many skeletons of this carnivore have been found in Mongolia. The largest specimen was over 45 feet long and 20 feet high.

Cenozoic era

LATER (MIOCENE
AND PLIOCENE)
PERIODS

26 million
years ago

First members of human family

Development of horned cattle

Development of frogs
and tailed amphibians

Cloven-hoofed animals

Lignite swamps and forests

Development of primates

Development of proboscids

Development of beasts of prey

Development of rhinoceroses,
camels, and tapirs

First primitive horses

Explosive development
of placental mammals

Great development of birds

Varied development of insects

Development of sea urchins,
crabs, and lobsters

Development of octopuses
and cuttlefish

Development of foraminiferans

Explosive development
of flowering plants

EARLIER
(PALEOCENE, EOCENE
AND OLIGOCENE)
PERIODS

65 million
years ago

Hyrachyus, an odd-toed ungulate that developed during the Eocene, was one of the ancestors of the rhinoceros and the tapir. In size, somewhere between a pig and a fox, it had a slim, horse-like body and was very agile. Its head was hornless and set on a short neck. Its four-toed forelimbs and three-toed hindlimbs were slender and fairly short.

Hyracotherium ("shrew-beast"), a small odd-toed ungulate about the size of a fox terrier, was the ancestor of the present-day horse. The evolution of the horse, which is well documented, started in the Eocene period, about 50 million years ago, and has continued to recent times. *Hyracotherium*, which lived in dense lowland forest, ate leaves and shoots. It had four toes on its front feet, and three on its hind feet; each toe ended in a hoof.

During the Eocene the grasslands of North America were inhabited by a flightless bird over nine feet high, known as *Diatryma*. It had massive limbs with three toes, a great head some 17 inches long, and a powerful hooked beak. In those days mammals were few and mostly rather small, so that birds as big as this probably represented a considerable danger to them. *Diatryma* lived in North America and Europe from the late Paleocene to the middle of the Eocene.

The ungulates that lived in North America during the Eocene include the *Uintatherium*, which was roughly the size of an African rhinoceros. The most striking feature of this and all related species was the shape of the skull, which tapered toward the snout and bore three pairs of bony, skin-covered "horns". Uintatheres were herbivorous, and the shape of their teeth suggests that they lived on soft juicy plants.

Basilosaurus, a primitive land mammal that became adapted during the early Cenozoic to living in the sea. A member of the order Cetacea, which today includes whales, dolphins, and porpoises, it evidently lived on fish, squid, and octopus. Sometimes over 80 feet in length, it had a fish-like body, and a low skull up to five feet long.

Below: All proboscids (mammals with trunks) are descended from *Moeritherium*, a small pig-like animal that appears to have originated in Africa about 40 million years ago during the late Eocene (early Oligocene) period. Its head was fairly long and the eyes were set well to the front, but it did not yet have a trunk. It lived in overgrown swamps and may have been partly aquatic.

Right: *Andrewsarchus*, which lived in eastern Asia during the late Eocene, represents an intermediate stage between mammals that specialized in a flesh diet (creodonts) and those that were vegetarian (condylarths). Both groups probably evolved from late Cretaceous insectivorous ancestors, although we know little of their origin. *Andrewsarchus*, which was about 13 feet long, was evidently omnivorous.

Left: *Orohippus* ("mountain-horse"), a more advanced representative of the horse line than *Hyracotherium* (p. 144), also lived in North America, in the middle of the Eocene, but in a different area. Its way of life was also different, for it inhabited dry grasslands scattered with bushes and lived on tougher vegetation than *Hyracotherium*, which it nevertheless still closely resembled in form and appearance. It stood about 16 inches high at the shoulder, but some of its foot bones had undergone further reduction, and striking changes had occurred in its teeth. The illustration shows a herd of *Orohippus* startled by the approach of a couple of huge uintatheriums (p. 146).

Brontotherium, a cumbersome animal that stood eight feet high at the shoulder, belonged to a group of odd-toed ungulates that lived in North America during the late Oligocene. Brontotherians lived on grass and possibly soft fruit. The snout was crowned with horny excrescences formed from the nasal bone. In some species the "horns" were large and branched; in others they grew as a single thick stem with a branched tip. The early Eocene ancestors of *Brontotherium* were about the size of sheep but their last, Oligocene, representatives were bigger than a rhinoceros.

Metamynodon (right) belonged to a group of primitive rhinoceroses that lived in North America at the same time as *Brontotherium*. It was hornless and grew to a length of about 15 feet. It had short, massive limbs and powerfully developed canines, so that at first sight it could be mistaken for a carnivore. The structure of the rest of its teeth shows, however, that *Metamynodon* was certainly herbivorous.

The huge *Arsinoitherium* was a unique type of primitive ungulate. About 10 feet long and almost 6 feet high, it resembled a rhinoceros in some respects, but kept all five primitive toes. Its chief characteristics were two large, thick nasal horns. fused at the

base, and two smaller horns on its frontal bone. *Arsinoitherium* was a sluggish swamp dweller and probably had habits similar to those of the present-day hippopotamus. Its remains have been found in lower Oligocene strata in Egypt.

Toward the end of the first half of the Tertiary era, Asia was inhabited by huge mammals that were utterly unlike any other members of this class, whether extinct or surviving. They included a gigantic hornless relation of the rhinoceroses, *Indricotherium*, the ancestor of the biggest land mammals ever known. The heavy body — about 27 feet long and 18 feet high — was supported by four massive three-toed limbs. Remains of this extraordinary animal, which was herbivorous, have been found in China and Mongolia.

Fossil remains of proboscids provide a fairly clear picture of the evolution of this group. *Gomphotherium* ("wedge-shaped beast"), one of its members, was distributed over North Africa, Europe, and eastern Asia during the Miocene; toward the end of the period it also appeared in North America. It was a long-jawed mastodon (mastodon meaning "nipple-toothed"—from the nipple-shaped cusps of the molars) with four slender tusks. In later forms the jaw became shortened and the lower tusks disappeared. *Gomphotherium* was the ancestor of many other proboscids, some of which became specially adapted to particular environments. It became extinct in the earliest part of the Pleistocene.

Amphibians abounded in the Tertiary swamps and lakes. Among them was *Palaeobatrachus*, a frog that resembled the African clawed toad. In some parts of central and western Europe, freshwater sediments are full of their skeletons. Sometimes they are so splendidly preserved that we can actually distinguish the outlines of their soft parts—internal organs and muscles—even the imprints of nerves, blood vessels, skin glands, and traces of pigment. Their tadpoles and eggs are sometimes preserved. Paleobatrachids are either the direct ancestors of recent frogs, or are closer to their ancestors than any other known group.

Miocene landscape, with deciduous trees and conifers. In the foreground are some of the first deer, of the genus *Palaeomeryx*, which were still without antlers, and members of the genus *Dicrocerus*, the males of which already had simple antlers. Flamingoes of the genus *Palaelodus*, whose beaks were less curved than in present species, can be seen feeding in a pool covered with water lilies of a species now confined to the Far East. A pair of *Gomphotherium* (p. 155) are browsing in the middle distance.

In the later part of the Tertiary, the plant life of central Europe was much like that of the subtropics today. Typical representatives of the early Miocene flora included birches (background, left) and tall conifers, bulrushes (front) and palms with wide,

fan-like leaves. Much of the landscape was occupied by great oaks, which were
represented by many different species.

Tetralophodon ("four-ridged tooth"), a typical mastodon of the Miocene and Pliocene periods, lived in Europe, Asia, Africa, and America. Its skull and neck, and particularly its lower jaw and tusks, were shorter than in *Gomphotherium* (p. 155), while its straight upper tusks were still quite long. It needed a trunk longer in proportion than those of its predecessors to enable it to reach the ground.

Among the many types of mastodon in Tertiary times was *Platybelodon* ("flat-tusk toothed"), which had incisor-tusks shaped like shovels attached to its long, curved lower jaw. It probably used them for rooting in soft, marshy soil. *Platybelodon* lived in Asia during the late Miocene, and in North America in the Pliocene.

Right: *Mesohippus* ("middle horse"), a new type of horse, appeared in the early Oligocene on the dry grassy steppes of North America. About the size of a prairie wolf, it had three toes on both front and hind feet, and was completely herbivorous.

Merychippus marks the next stage — after *Mesohippus* — in the evolution of the horse. It already had limbs similar to those of present-day horses but they were shorter. Like *Mesohippus*, it had three toes on both front and hind feet, but the central toe was much stronger than the two side toes, which did not reach the ground. About as big as a donkey, it lived from the middle of the Miocene to the early part of the Pliocene.

Teleoceras, an unusual type of rhinoceros with very short legs, had a single thick horn mounted on its nose. In general appearance it resembled a hippopotamus rather than a rhinoceros; indeed, it may well have lived at least partly in the water. *Teleoceras* is found in North America from the Miocene period to the late Pliocene.

Vast swampy forests were a marked feature of the Tertiary era, appearing in many of the low-lying land areas that emerged as a result of extensive mountain-folding. To give themselves additional support, conifers grew long prop roots, while the lower trunks of deciduous trees became cone-shaped. As the original carbon-containing material of the forests became buried, it was subjected to enormous pressure. With Tertiary swamp deposits, the carbonization process only went to the stage of lignite, or brown coal. This contains much less carbon than the earlier Carboniferous deposits, which themselves vary in grade from bituminous coal to anthracite.

Fossil remains of the giant salamander, *Andrias scheuchzeri* (below), were discovered long before zoologists had found it as a surviving form in eastern Asia. It is known in Europe from Oligocene and Pliocene strata, in North America from the Miocene to the Pliocene, and in eastern Asia from Pleistocene to recent times. ("Andrias" in Greek means "man-image" or "statue", and a fossil discovered in Germany in 1726 was originally thought to be the remains of a man drowned in Noah's Flood. It was therefore called *Homo diluvii testis*— "Man, a witness to the Flood". Only later was it shown to represent a salamander, whence the generic name.)

Thylacosmilus ("pouch-knife"), a carnivorous marsupial, flourished in South America during the Miocene. It had long teeth like those of the sabre-toothed tiger (p. 172).

During the Tertiary some of the flightless birds—such as *Phororhacos*, which was 10 feet high—reached huge dimensions. *Phororhacos'* head was longer than that of a fully grown horse. It devoured reptiles and small mammals. The great claws and hooked beak testify to its carnivorous habits. It lived in South America in the plains and foothills of Patagonia, where it was probably able to develop only because placental beasts of prey were then absent.

Opposite: Proboscids of the genus *Deinotherium* are an isolated and, from the evolutionary point of view, a still obscure group. The front of the lower jaw curved downward almost at right angles and supported a pair of tusks of which the use is not clear. The upper tusks were not developed at all. *Deinotherium*, which grew to a height of about 13 feet, lived in Europe from the early Miocene to the late Pliocene. Its remains have also been found in Africa and Asia.

Proconsul africanus (below), an early primate that lived in Kenya during the Miocene period, may be very like the common ancestor from which both the modern apes and modern man must derive. Its remains were discovered in the 1930s. Although *Proconsul* is possibly related to the chimpanzee, its skull lacks the great brow ridges above the eye. The structure of the several skeletons found shows that these animals were still quadrupeds.

A herd of mastodons of the species *Anancus*, which was quite common in Europe and Asia from the early Pliocene to the Pleistocene, and also in Africa during the Pleistocene. In this species the upper tusks were up to 14 feet long.

The sabre-toothed tiger, *Machairodus*, seen here attacking a small horse, belonged to the cat family (Felidae), but was not a true cat (subfamily Felinae), like the lion or the tiger. In the true cats, the upper canines grew smaller during evolution, while in the sabre-toothed group they grew larger. Sabre-toothed tigers evidently hunted large animals like gazelles, antelopes, and wild boars. They reached their highest degree of specialization at the beginning of the Pleistocene and then, soon afterwards, they seem to have died out.

HOLOCENE
PERIOD

10,000
years ago

Quaternary era

Modern man

Mammoth hunters

Neanderthal man

Cave bear

Cave lion

Aurochs

Woolly rhinoceros

Mammoth

Sabre-toothed tiger

Giant Red Deer

Homo erectus

Australopithecus

Development of hominids

PLEISTOCENE
PERIOD

2 million
years ago

Development of rodents

Taiga landscape during the Quaternary, an era that was characterized by marked cooling and by spreading sheets of ice. The taiga, which may have extended 200 miles or more from the edge of the northern ice sheet, contained conifers and some deciduous trees. The most characteristic animal was the bear.

A herd of forest elephants *(Palaeoloxodon antiquus)*, which lived in Europe, Africa, and Asia during the interglacial periods of the Pleistocene. A huge animal, bigger than a mammoth, *Palaeoloxodon antiquus* stood 14 to 17 feet high at the shoulder and had straight tusks that were often more than 10 feet long.

Various pygmy forms of elephant, hippopotamus, deer, and dormouse evolved in certain areas of the Mediterranean region during the early Pleistocene. When the islands of Malta, Crete, and Cyprus were formed, the animal populations thus isolated lived under new, restricted conditions, and eventually developed many pygmy forms. In the late Pleistocene, when Malta was completely cut off, a local form of elephant—*Loxodonta falconeri*—evolved. Only three feet high, it was no larger than a pony.

Mammoths had curved tusks, which were sometimes crossed in old males. *Archidiskodon meridionalis*, seen here, was related to the Indian elephant, but in body structure it was more like the African; its tusks curved slightly upward. It reached a height of over 12 feet and lived in Europe, in the Mediterranean region, in the early and middle Pleistocene.

Left: *Paranthropus* ("almost-a-man") *robustus*, whose remains were found at Krom-draai, near Johannesburg, South Africa, in 1938, by Robert Broom, a South African paleontologist. *Paranthropus* lived between three million and one million years ago, at the end of the Tertiary and the beginning of the Quaternary. About five feet tall, he weighed around 150 pounds. He was biped, but did not walk absolutely erect. He had a more massive skull than *Australopithecus* (right), a large lower jaw, and a thick ridge of bone on which the muscles of mastication were inserted. *Paranthropus* was so well adapted to forest living and a vegetarian diet that further development was impossible, and he had no successor. In recent years, paleontologists have tended to abandon the generic name *Paranthropus*, and now classify this find as a member of the earlier genus *Australopithecus*.

Above, right: *Australopithecus africanus* ("southern ape of Africa"), whose remains were found near Taung, 80 miles north of Kimberley, Botswana, in 1924, by Raymond Dart, another South African paleontologist. A contemporary of *Paranthropus* (left), he was about the size of a chimpanzee or a young gorilla, and walked erect. The structure of his teeth shows him to have been omnivorous, but largely flesh-eating. *Australopithecus* was a steppe-dweller and a hunter, so his environment and mode of life were very different from those of *Paranthropus*. Scientific analysis of his anatomy showed that he was closer than any other creature to the hypothetical ancestor of modern man. *Australopithecus*, who is shown opposite holding a stick, must have been able to use such unadapted implements when hunting.

In the early Pleistocene period, the Etruscan rhinoceros was the only representative in Europe of *Dicerorhinus*, the genus to which most European ice-age rhinoceroses belonged. The genus—which still has a living representative in Sumatra, Indonesia—had two horns, unusually long legs, and perhaps a somewhat hairy hide. It was over eight feet long and five feet high, and lived on grass and foliage in a forest environment. These animals were common in Europe and Asia in the late Tertiary.

Mastodon americanus, the most familiar American mastodon, reached a length of 12 feet, had long tusks curving slightly upward and outward, and had a short, high skull. Unlike European species, its skin was covered with hair, which has often been found preserved with the skeleton. Mastodons probably evolved during the Miocene, and before that period ended their representatives had spread to America. In North America, *Mastodon americanus* survived into the Pleistocene; and there is no doubt that it was a contemporary of man.

Mammuthus imperator, one of the biggest proboscids, lived on the southern Great Plains of North America during the middle of the Pleistocene period, at the time of the First Interglacial. It was over 12 feet high and so taller than the African elephant today. Its short, flat-backed skull had a prominent boss on the crown, and its huge

tusks were up to 14 feet long. Well-preserved skeletons have been found in Texas, Colorado, Nebraska, Kansas, and Oklahoma, as well as in the tar pits at Rancho la Brea, now a public open space in Los Angeles, California.

Mammuthus columbi, seen here in an encounter with the North American sabre-toothed tiger *Smilodon* and its young, lived during the Pleistocene in what is now the south-eastern part of the United States. It was particularly common in Florida, Georgia, the Carolinas, and Louisiana, and extended as far south as Mexico. It had a lower, longer skull and a rounder crown than other mammoths.

Mammuthus trogontherii may have been the direct ancestor of *Mammuthus columbi* (opposite) in North America. Bigger than any other European mammoth, it stood 15 feet high at the shoulder; the largest pair of tusks found measured about 17 feet. As the picture shows, it was probably somewhat more hairy than earlier species. Excellently adapted to life on the grass steppe environments of Europe and Asia, it lived during the later Pleistocene.

The sword-toothed "cats" (Homotheriini) are classified as a special "tribe" of the sabre-toothed subfamily, Machairodontinae (p. 172). Homotherians, which were very common in Europe, differed from the sabre-toothed tigers in several ways. Their canine teeth, though long, were shorter than those of *Smilodon* (p. 190) and were more curved and flat, like thin scimitars. The curved form shows that *Homotherium* ripped its prey to pieces. The disappearance of these beasts of prey is perhaps associated with the disappearance of the large pachyderms ("thick-skinned"—elephants, rhinoceroses, hippopotamuses), which may have been their main prey.

The great fossil rhinoceros *Elasmotherium sibiricum*, which lived on the steppes of southern Russia but occasionally moved into central Europe, was the biggest rhinoceros ever known. Its horn was about six feet long. The first specimens were found in Spain (Miocene) and others in China (Pliocene). *Elasmotherium* flourished in Russia during the Pleistocene.

Opposite: The great sabre-toothed tiger *Smilodon californicus*, which lived in North and South America during the Pleistocene, about to attack *Nothrotherium*, a primitive ground sloth about the size of a pig. *Nothrotherium* was a slow, ungainly animal; its feet were armed with large, curved claws and it was able to stand on its hind legs and pluck foliage from the trees. In the Argentinian pampas, *Smilodon* was probably largely responsible for the disappearance of various types of herbivorous animals, including *Nothrotherium*. *Smilodon neogaeus* (above), another sabre-tooth, reached South America only in the Pleistocene, after the northern and southern parts of the continent had been joined by a land bridge.

Left: *Homo erectus modjokertensis*, whose remains were discovered at Modjokerto, Java, in 1936. This creature was about five feet six inches tall and already used tools. Not only did he have a larger brain than *Australopithecus* (p. 180), but his skull was more massive, with large brow ridges, receding forehead, prominent jaws, and no chin. He lived between one million and 500,000 years ago.

Right: *Homo erectus heidelbergensis*, whose jaw was discovered in the Heidelberg district of West Germany in 1907 and 1908, lived about half a million years ago. He is the oldest fossil man yet found in Europe. The teeth are human and the canine points stand no higher than the rest of the tooth-row. Like Java man and Peking man, Heidelberg man may have learned to use simple tools.

Below: A reconstruction showing how a group of Heidelberg men may have looked.

Madagascar, an island off the southeast coast of Africa, has been separated from the mainland for many millions of years. Its fauna is both peculiar and unique. Giant "elephant" birds, for instance, reached the peak of their development during the Quaternary era, when they grew to a height of about 10 feet. They resembled ostriches and their best known representative, *Aepyornis maximus* (below), lived only in Madagascar. The birds died out in about the 17th century, when the island's virgin forests were destroyed.

New Zealand, like Madagascar, is an oceanic island-group with a peculiar fauna, which had developed unmolested by man or carnivores. Moas, great ostrich-like birds, lived there until about the middle of the 17th century. They were about 13 feet tall and had a sturdy skeleton, powerful legs, a small skull, and a short, flat beak, which was sometimes curved, sometimes sharp and pointed. The wings and shoulders were undeveloped. Several genera and species of moa are known, the largest being *Dinornis maximus*, illustrated here.

The emu family, which is related to the moas, was also well represented in New Zealand. These emus were large, robust birds with many features in common, but there were differences in leg-bone structure between species. *Emeus crassus* (below), which lived in the South Island, had a short, straight beak curved at the tip and a smaller body than its cousins. In this bird, the bone of the shank was always thicker than that of the thigh.

Euryapteryx was another flightless bird belonging to the emu family. It was a massive animal with a blunt skull and a short, wide beak. One species, *Euryapteryx elephantopus* ("elephant-footed"), is depicted above. Our knowledge of moas is fairly complete, because many of their bones, together with the remains of soft parts of their bodies, have been preserved. Footprints have also been discovered. Moas lived in scrub country and in forests at varying altitudes.

Homo erectus erectus of Java, formerly called *Pithecanthropus* ("ape-man"), had a small, wide nose, a receding forehead, large brow ridges, massive prominent jaws, and no chin. He was already able to make the most primitive stone tools, which in Java were

found in the same layers as his bones. These implements were chipped on one side only. *Homo erectus* also used sticks or clubs as primitive weapons.

Below: *Homo erectus erectus* probably led a simple nomadic life, moving from place to place in small troops in search of food. No doubt, he lived on the juicy fruit of tropical trees, on various edible tubers and roots, and by hunting for meat. He had to learn to defend himself against animal enemies as well as against the natural hazards of storm, typhoon, earthquake, and volcanic eruption—not uncommon in Java. He was well equipped for all these purposes, because he had long legs, so that standing upright, he had a clear view of his surroundings. In addition, the habitually erect posture freed his hands for activities other than mere movement from place to place. His hands and his now considerable brain were, in fact, his main assets in the struggle for existence. He was probably the first of our predecessors capable of constructive mental processes.

Peking man *(Homo erectus pekinensis)*—left—whose remains were found at Choukoutien, near Peking, between 1918 and 1937, had already discovered how to control and use fire. No one knows how man learned to do this, but his first introduction must have been through fires started by lightning, hot lava, forest fires, or other natural means. Making fire by artificial means, by friction or otherwise, probably came later. For Peking man, fire must have been a great asset in the cold winters, and the need to learn how to preserve and propagate it must have helped to sharpen his intelligence. He lived perhaps 220,000 years ago, rather later than Java man, whose date is estimated at between 500,000 and 400,000 years ago.

Below: Portraits, in reconstruction, of a Peking man and woman.

Of all the extinct mammals, none is more famous than the woolly mammoth (above). Some were bigger than present-day elephants, being almost 14 feet high. Woolly mammoths lived in Europe and the northern parts of Asia and America from the middle of the Pleistocene to its end. The last of them died out 10,000 years ago.

The huge woolly mammoth, *Mammuthus primigenius*, which flourished mainly during the Pleistocene, was a typical inhabitant of the tundra and an important member of the ice-age fauna. Whole carcasses have been found in many places in the permanently frozen soil of Siberia, and this has allowed us to study the soft parts and skin. From the stomach contents, we know that these mammoths lived on the young twigs of conifers, willows, birches, and alders and on different steppe plants. Food supplies were adequate only in summer, when the animals built up reserves, chiefly of fat, in the form of a hump on the shoulders. This enabled them to survive severe shortages when winter came and their sources of food were covered by snow. Woolly mammoths probably migrated southward for the winter and returned north again every summer.

R. Burian 61

A herd of woolly mammoths, in search of a meal, using their tusks to scrape aside the first snow of winter before migrating south to warmer climes and greener pastures. Much of our information about the mammoth comes from men who lived during the ice ages and drew or engraved its image on the walls of caves or on fragments of bone and tusks.

Neanderthalers *(Homo sapiens neanderthalensis)*, one of the main subspecies of man, probably evolved in Asia or eastern Europe about 200,000 years ago. The Neanderthal people were the authors of the flint industries (collections of implements) known to archaeologists as "Mousterian". Until recently, they were believed to be the immediate predecessors of modern man, but most authorities now regard them as an extinct subspecies of *Homo sapiens*.

Right: Remains of West European Neanderthalers (who lived at the start of the last glacial period) or of types bearing a general resemblance to them, have been found in different parts of Europe, Africa, and Asia. We have therefore a good idea of what these primitive humans looked like. Neanderthal man differed in many respects from the modern subspecies, *Homo sapiens sapiens*. His usually large and receding skull had huge prominent brow ridges and wide cheekbones, on which some of the powerful muscles of mastication originated. His nose was broad at the root and he had no chin; but his teeth were typically human. He had a stockily built body, with a short neck. The average male stood about five feet six inches tall.

Like many other groups of Pleistocene animals, the deer displayed a tendency to produce gigantic forms. The giant deer *Megaloceros giganteus* (shown here) appeared in the Old World. It was a powerful beast with an antler span of up to 14 feet. Since its palmated antlers resembled those of the present-day elk and fallow deer, it was once wrongly regarded as a giant form of one of these species. It lived in wide, open plains with grass and shrubs, and probably avoided forests, where its antlers would have been a hindrance.

The cave bear *(Ursus spelaeus)* is one of the most familiar animals of the ice ages—thousands of its bones have been found in European caves. Often enormous, it was much larger than the brown bear. It had a head large in proportion, with a bulging forehead and short, sturdy legs, and was usually about the size of a grizzly. It had a life span of about 20 years and must have been largely vegetarian. Local dwarf forms developed high up in Alpine caves (above 8000 feet) and in the Harz mountains in Germany. The cave bear was sometimes hunted by man (p. 218), but the piles of bones found in some caves show that these animals often died a natural death.

The woolly rhinoceros, *Coelodonta antiquitatis*, was a regular companion of the mammoth and a contemporary of prehistoric man, who often hunted and drew it. It reached Europe, presumably from Asia, in the middle of the Pleistocene (during the Third Glacial), and died out only 10,000 years ago. It had two horns, and a long, hairy coat to protect it from the cold. Unlike the temperate-forest rhinoceros, it walked with its head and neck held low.

Panthera spelaea, the cave lion, which appeared in Europe in the middle of the Pleistocene, was probably the biggest feline beast of prey ever to exist. It was about one
third larger than an African lion today. Specialists are still undecided whether it
was a lion, a tiger, or a separate species in its own right.

The mammoth-hunters pictured on these two pages were among the earliest representatives of *Homo sapiens sapiens*, having appeared rather suddenly in Europe some 30,000 to 40,000 years ago. The many marks of their presence, which include skilfully made tools, traces of hearths and dwellings, and the remains of animal prey, have revealed many details of their habits and culture. The often large quantities of bones found near his living-places (usually caves) show that early man generally hunted in groups; he was incapable of tackling very large animals single-handed. His most important game was the mammoth, which could provide a large supply of food for a long time. Central European hunters like those shown here knew from experience that the mammoth migrated north in the spring and returned south before the onset of winter.

Left: Chosen scouts keeping watch on the mammoth's regular migratory routes are ready to give the signal to the main party of hunters.

Above: The hunters have broken up the mammoth herd, isolated one of the younger and less experienced animals, and driven it into a carefully concealed trap (probably in a swamp) from which there is no escape. While some of the party throw rocks to divert the beast's attention, others—armed with spears and similar weapons—move in for the kill.

Last rites for a mammoth-hunter. The grave of this prehistoric man was found 15 feet below the surface, in an undisturbed layer of loam, at Brno, Czechoslovakia, in 1891. The dead man must have been greatly esteemed by his companions, for the shallow grave was lined with furs on which his body, dressed in skins, was then laid. Round his neck he had a tusk-shell necklace, and beside him lay two flat limestone disks with a central hole, three small stone disks with perforated edges, and 11 smaller platelets of mammoth bones. In addition, there were weapons, and implements made of ivory and stone. The body had been sprinkled with red ochre and was covered with a mammoth's shoulder blade, supported by a tusk. Gifts to the dead man included the rare figure of a naked man carved from a mammoth's tusk.

Below: The "Old Man of Cro-Magnon", a reconstruction based on one of five Stone Age skeletons found in 1886 beneath an overhanging wall of rock known as Cro-Magnon, in the Dordogne region of southern France. One of the skeletons, which was so-nicknamed, acquired greater fame than the rest and was adopted as the prototype of the "Cro-Magnon race". The male members were often over six feet tall and well built; their skulls were generally of medium length, with strongly developed, but never Neanderthal-like, brow ridges.

Below: An artist of the late Stone Age (some 20,000 to 30,000 years ago) at work on the figure of a woman, known as the Věstonice Venus. To make the figure—which was found in Dolní Věstonice, Czechoslovakia— the artist used a new material prepared by mixing ash with bone powder, clay, and fat. When baked, this became stone-hard. Marks found on other figures and on lumps of this material show that the fingerprints of this prehistoric artist differ in no way from those of a modern man.

Above: Stone Age people often buried their dead near the camp. Some tribes put the body in a shallow grave, with stones round the head and feet; others weighted the head, chest, and legs with heavy stones, presumably to make sure the dead would not rise. Some bodies were bound hand and foot, and placed in the grave with their knees drawn up; others were simply left in a cave and the entrance was then blocked with stones. Weapons, gifts, ornaments, and food were usually laid by the corpse.

Even when using their long spears, hunters of the Old Stone Age must have found the cave bear (p. 209) a formidable opponent.

The first humans appeared in northern Europe about 25,000 years ago, once the ice sheet had retreated northward and the country of what is now northwest Germany had developed a tundra vegetation. These men were reindeer hunters, their main weapons being harpoons, and bows and arrows. They produced very few works of art—a handful of engravings and decorated objects is all that has been found. The picture on the left shows a typical reindeer hunter.

Settled farmers of the New Stone Age appeared in central Europe and in the Danube valley about 3000 B.C. Their implements were primitive and their fields were small. Wheat, barley, millet, beans, and peas were the main crops. They also kept domesticated animals and made beautifully decorated pottery The man pictured here is sharpening his "polished" stone axe, one mark of a Neolithic stage of economy.

The science of taxonomy (from a Greek word meaning "arrangement") deals with the classification and naming (nomenclature) of plants and animals, and hence with the relationships that exist between living things. All living things are assigned to one of two kingdoms, plant or animal, which are then subdivided into smaller, or lower groups, forming a system based on increasing similarity as we descend the hierarchy:

Kingdom
Phylum (plural: phyla; from a Greek word meaning "tribe"). The term is generally
       reserved for animals, the corresponding term for plants being Division.
Class
Order
Family
Genus (plural: genera; from a Greek word meaning "race")
Species (from a Latin word meaning "appearance")

For instance, the domestic cat belongs to the kingdom Animalia; to the phylum Chordata (animals with a straight spinal cord and with bilateral symmetry); to the class Mammalia (chordates that bear their young alive and suckle them); to the order Carnivora (mammals that are flesh-eaters); to the family Felidae (including cats, great and small); to the genus *Felis* (small cats); and to the species *catus* (the domestic cat).

Parallel with the phylum Chordata are numerous phyla of animals without backbones, ranging from the single-celled microscopic Protozoa ("first animals") to highly organized multicellular Arthropods ("joint-footed"), including lobsters, insects, and spiders, often with armoured external skeletons. In between come the vast multitudes of different worms, jellyfishes, sea urchins, and shellfishes, to name only a few of the more familiar.

Classes included in the phylum Chordata, alongside the Mammalia, are devoted, for instance, to fishes, reptiles, and birds. To make possible even more subdivisions, zoologists have created subclasses, infraclasses, superorders, suborders, superfamilies, subfamilies, subgenera, and subspecies—as well as cohorts and tribes—as seemed necessary.

Nomenclature

The Swedish botanist Carl von Linné (1707—78), better known by his latinized name of Carolus Linnaeus, introduced the binomial ("two-name") system of nomenclature, by which any known plant or animal could be accurately identified by a pair of names, one for the genus (the generic name) and the other for the species (the trivial name); the genus and the trivial together make up the specific name.

These two names are treated as if they were Latin, and are always written in italics; the generic name is capitalized, but the trivial name is regarded as adjectival and always has a small letter. Names of families and higher groupings are also given a capital letter, but are not italicized. Families are identified by the endings -idae, superfamilies by -oidea, and subfamilies by -inae. Higher groups than these have no systematic termination.

Nomenclature is governed by a system of international rules, the most basic of which is that the trivial name published with the first adequate description can never be altered. The generic name, on the other hand, may be changed with due cause if desired.

Trinomials (three names) are nowadays sometimes used to distinguish subspecies or races. Examples used in this book include the names *Homo erectus erectus* (the "type"-race) from Java and *Homo erectus pekinensis* (Choukoutien, China).

The simplified classification set out on the next two pages shows only the higher divisions of plants and animals. But with reasonable practice, the reader will find it easy to see where, for instance, a particular species belongs and how it may relate to other groups of living, or once living, things.

Kingdom: Plantae

Divisions		
	Bacteriophyta	Bacteria
	Cyanophyta	Blue-green algae
	Myxophyta	Slime-moulds
	Euglenophyta	Mobile aquatic plants
	Chrysophyta	Yellowish algae, diatoms
	Phyrrhophyta	Including dinoflagellates (unicellular plants)
	Chlorophyta	Green algae (mainly of fresh waters)
	Phaeophyta	Brown algae, including shore seaweeds
	Rhodophyta	Red algae, deeper marine weeds
	Fungi	Three classes: moulds, yeasts, mushrooms, etc.
	Lichenes	Fungus and algae, symbiotic
	Bryophyta	Two classes: liverworts and hornworts, mosses
	Pteridophyta	Four classes, including club mosses, horsetails, and ferns
	Gymnospermae	Four classes, including cycads, ginkgoes, seed-ferns, conifers, etc.
	Angiospermae	Flowering plants, seeds contained in a vessel. Divided into two classes:

Classes of Angiospermae	
Monocotyledoneae (with a single seed leaf, leaves parallel-veined)	Ten orders, including palms, lilies and irises, grasses and sedges, orchids
Dicotyledoneae (with two seed-leaves and net-veined leaves)	Thirty orders, including most familiar forest trees, wild and garden flowers, vegetables, and fruit trees

Kingdom: Animalia

Subkingdoms		
	Protozoa (single-celled animals)	Two subphyla, five classes
	Metazoa (many-celled animals)	Numerous phyla, including:

Phyla		
	Parazoa	Sponges, some with hard parts
	Archaeocyatha	Short-lived group of sponge-like animals in early Cambrian
	Cnidaria	Coelenterates, having a digestive cavity with a single opening and radial symmetry. Two forms: medusa (e.g. jellyfish) or polyp (e.g. sea anemone, corals)
	Ctenophora	Comb-jellies, sea-gooseberries, Venus' girdle, etc.
	Platyhelminthes	Flatworms, including flukes and tapeworms (parasitic)
	Bryozoa	Sea mats and fans, important in reef-building
	Brachiopoda	Lamp-shells, in the geological past very successful, nowadays much reduced in numbers and variety
	Annelida	Segmented worms, largely marine
	Arthropoda	Invertebrates having a segmented structure with a hard external skeleton of chitin and many-jointed limbs (e.g. crabs, lobsters, barnacles, spiders, mites, insects, centipedes). Six subphyla, several with a number of classes and subclasses, many orders and multitudinous lower groups, many extinct (e.g. trilobites and eurypterids). An enormously successful phylum

Phyla

Mollusca — Shellfish—univalves, bivalves. Six classes in all, including Cephalopoda—"head-footed", with tentacles surrounding the mouth and an internal shell (e.g. squids, octopuses, cuttlefish, and many extinct orders)

Chaetognatha — Arrow-worms

Pogonophora — Tube-worms

Echinodermata — "Spiny-skinned", generally with five-part radial symmetry—crinoids, sea cucumbers, sea urchins, starfishes. Many groups extinct

Hemichordata — Tongue or acorn worms

Chordata — Bilaterally symmetrical animals with a notochord (rod of gristle along the back) and nerve-chord internal axis. Four subphyla, including:

Subphylum

Vertebrata — Chordates with an internal skeleton of cartilage or bone:

Classes of Vertebrata

Agnatha   Jawless vertebrates (e.g. ostracoderms, lampreys)

Placodermi   Vertebrates with primitive jaws. All extinct

Chondrichthyes   Cartilaginous fishes (e.g. sharks, rays)

Osteichthyes   Bony fishes. Two subclasses, several superorders and orders, many primitive or extinct or represented only by very few—or even a single *(Latimeria)*—living species. Superorder Teleostei includes most numerous and successful living fishes

Amphibia   Breed in water and have a fish-like gilled larva (tadpole). Adult air-breathing, with lungs. Two subclasses, several superorders and orders, including the extinct labyrinthodonts (ancestral to the reptiles), living frogs and toads, newts and salamanders

Reptilia   Cold-blooded, mainly terrestrial. Lay eggs. Often covered with scales. Six subclasses and numerous orders, mostly extinct. In the Mesozoic the most successful class of vertebrates, often of huge size and very numerous. Represented today by only four living orders: Chelonia (tortoises, turtles), Squamata (lizards, snakes), Crocodilia (crocodiles, alligators), and Rhynchocephalia (*Sphenodon*, of New Zealand, the sole surviving genus and species)

Aves   Birds. Consists of two subclasses: 1. Archaeornithes—extinct *Archaeopteryx* the sole member. 2. Neornithes—all other birds; divided into 29 orders, some, like the huge flightless birds of Madagascar and New Zealand, only recently extinct

Mammalia   Three subclasses and three infraclasses, mostly extinct, but include the very primitive Monotremes (e.g. *Ornithorhynchus*, the duck-billed Platypus) and Marsupials (e.g. *Didelphis*, the Australian opossum). The third infraclass, Eutheria (true mammals) includes all the familiar mammals besides, grouped in four cohorts, comprising 25 orders, of which a few are entirely extinct, while most have many extinct, beside the surviving, members. Widespread orders today are those of the insectivores (shrews, moles, hedgehogs, etc.), bats, primates, rodents, carnivores, proboscids, odd-toed and even-toed ungulates. The primates, including tarsiers, lemurs, monkeys, apes, and man, fall noticeably among the structurally primitive and unspecialized, in all respects save their intelligence and adaptability. As in time, so in classification: the mammals and man figure only on the last couple of pages of a very long book, and are vastly outnumbered by the multitudes of other living things.

Glossary

algae Several distinct botanical divisions of primitive plants, of sea and fresh water. Many are single-celled and microscopic, some are very large.

ammonite Extinct cephalopods (q.v.) with a generally spiral, many-chambered shell. Named after the similarly shaped "horn of Ammon", the ancient Egyptian ram-headed god.

amphibian "Living in both" elements —water and air. Moist-skinned animals (e.g. frogs, newts, and salamanders) that have to breed and lay eggs in water and go through a fish-like stage (tadpole), breathing at first by gills, though adults are air breathing and may leave the water temporarily.

angiosperms The botanical division of green flowering plants with seeds growing in an ovary.

Archaeozoic The earlier and larger part of the Precambrian (oldest) geological era, in which there is only very slender evidence of life-forms on earth.

arthropods Animals having many-jointed external armour or chitinized skin-plates and numerous (10, 8 or 6) legs. Lobsters, spiders, insects are included.

Azoic The biologist's term for the geologist's Archaeozoic (q.v.)

bipedal An animal that uses only two limbs for walking.

bivalve A soft-bodied animal with two parts to its shell, hinged together.

brachiopods Bivalve animals classified in a phylum distinct from that of the superficially similar mollusks. Commonly known as "lamp-shells" because the larger shell has an opening like the wick-hole in a Roman oil lamp.

Cambrian The first period of the Paleozoic era. Named after Cambria (the Roman name for Wales), where the rocks belonging to the fossils were first studied.

Carboniferous The fifth period of the Paleozoic era. The name means coal-bearing since most coal-yielding strata at first known were formed within it. Not all Carboniferous rocks contain coal. American geologists prefer to divide the period into two systems, the Pennsylvanian (upper) and the Mississippian (lower).

carnivore A flesh-eating mammal—a member of the order Carnivora.

Cenozoic The latest era of living things, ranking with the Paleozoic and Mesozoic eras.

cephalopods Members of the class Cephalopoda, which includes squids, octopuses, and cuttlefishes. Highly mobile advanced carnivorous mollusks with a soft bag-like body and a series of tentacles armed with suckers surrounding the mouth. They have well-developed eyes and nervous systems.

chalk A soft marine limestone of Cretaceous age, often rich in fossils, mostly skeletons of minute animals such as foraminiferans.

chordates Members of the phylum Chordata, which includes all animals with backbones (vertebrates) as well as some less well known that share the same basic structure of a longitudinal nervous axis and bilateral symmetry, but have a stiffening axis of cartilage (gristle), not bone.

coelenterates A phylum of animals having radial symmetry and a body cavity with only one opening, by which food is taken in and digested and the waste ejected. Familiar members are jellyfishes and sea anemones.

cotyledon The "seed-leaf" of a flowering plant's embryo. It serves as a store of nourishment to sustain the developing embryo after germination and until it has developed a root-system of its own. Flowering plants are divided into two classes according to whether their seeds contain a single cotyledon or a pair.

cotylosaurs A group of primitive reptiles of late Paleozoic and Triassic age, which gave rise to more specialized types in the succeeding two periods of the Mesozoic.

creodonts Primitive mammals of the early Tertiary from which the present-day carnivores are probably developed.

Cretaceous The final period of the Mesozoic during the latter part of which the chalk (Latin *creta*) was formed.

crustacean A member of a class in the phylum Arthropoda in which the body has an armoured, external, many-jointed skeleton of chitin, a substance with horn-like properties. Among common crustaceans are crabs, lobsters, and woodlice.

cusp The point, or one of the points or prominences, on the crown of a tooth.

cycads Members of a class of the gymnosperms (q.v.), that division of the plant kingdom that includes the seed-ferns and ginkgoes, with long fern-like leaves. Many were large and tree-like in habit. Cycads were more widespread in the Mesozoic, but have few representatives today. They have naked seeds, borne generally in a large central cone.

Devonian The fourth period of the Paleozoic era. Rocks laid down at this time were first distinguished in the county of Devon, England, but their occurrence is world-wide.

dicotyledons One of two main classes of flowering plants characterized by the presence of two cotyledons (q.v.) in the seed. Dicotyledons have relatively broad leaves with a branching system of veins. Most common trees and herbs in the temperate zone belong to the class.

dinosaur A rather loose name for a group of often very large Mesozoic land reptiles. The group includes two structurally distinct types — the Ornithischia, of which the hinder parts of the pelvis resemble those of birds, and the Saurischia, which in this respect are closer to lizards. The former were exclusively vegetarian, sometimes defensively armoured, the latter included both carnivores and herbivores.

dorsal The backward aspect or view of a structure, or its backward position in relation to others. It is independent of bodily attitude. The dorsal aspect of a man standing is from *behind* him; of a standing horse or other quadruped from *above*.

echinoderms A phylum of marine animals that includes the sea urchins and starfishes, marked by generally five-part radial symmetry, an internal limy skeleton, a spiny or warty surface, and water-filled tube-feet for locomotion.

environment The sum of the factors, natural or artificial, that control the surroundings and conditions of life, of a plant or animal. In paleontology, evidently, the term implies the *natural* surroundings. Man, interfering with nature, creates for himself a largely artificial environment.

Eocene The second period, beginning some 54 million years ago, of the Tertiary era. With the extinction of most of the reptiles at the end of the Cretaceous, the mammals underwent explosive evolution in the Tertiary, a process that began in the Paleocene.

era A major division of geological time (cf. period).

eurypterids Members of a long extinct (Ordovician, Permian) subclass of marine arthropods (q.v.), often equipped with long oar-like swimming legs; they were probably carnivorous predators.

evolution The process by which living things develop from simpler, generalized forms to the more complex and specialized, each generation being better fitted to survive and propagate in the particular niche of the environment it occupies. When the environment changes, the too highly specialized types are unable to adapt and so become extinct.

fauna A community of animal species living together in a particular environment, area, or period.

flora A community of plant species found together in a particular environment, area, or period.

gastropods Univalve mollusks (snail-like) with a flat extensible "foot" for crawling about and a generally spiral protective shell. Examples to be found today include: whelk (marine), ramshorn snail (freshwater), both with gills; Roman snail slug (terrestrial, air-breathing).

genus (plural: genera) The next higher taxonomic group to the species in botanical or zoological classification. Thus a genus consists of one or more species, and one or more genera are grouped together in a family.

glaciation The development over a land mass of an ice sheet, such as exists today in Greenland and Antarctica. Such inland ice sheets covered much of northwest Europe and North America at several times during the last million years.

gymnosperms Members of the botanical division Gymnospermae, green vascular plants with only one seed-coat, the seeds generally borne in a cone. Examples include cycads (q.v.), conifers, cordaits, and ginkgoes.

herbivore A vegetarian mammal that feeds (grazes) mainly on grasses and herbs in open country rather than on the foliage of woody shrubs or trees, which are preferred by many woodland plant feeders (browsers). Many species, of course, both graze and browse when the opportunity offers.

Holocene The latest geological period (also called Recent or, in glaciated regions, Postglacial) in which all the animal species are of the modern type, though some are extinct. In terms of time, the Holocene corresponds to the last 10,000 years.

hominid A member of the zoological family Hominidae, which includes the genera (q.v.) *Homo* and *Australopithecus.*

ichthyosaur A marine carnivorous reptile, fish-shaped, with flippers derived from a reptilian forelimb, and whale-like flukes to its tail.

invertebrate An animal without a backbone. Invertebrates long preceded the first Ordovician vertebrates and their species vastly outnumber the vertebrates today.

Jurassic The middle of the three periods of the Mesozoic era. Rocks and fossils attributed to it were first named from the Jura mountains on the border of France and Switzerland.

mammal A warm-blooded vertebrate, member of the class Mammalia. Distinguished by a four-chambered heart and a coat of hair, it bears its young alive and suckles them in infancy with milk from paired glands on the ventral surface of the body.

marsupials Primitive mammals (most today confined to Australasia) that bear their young, after very brief gestation, in an almost embryonic, extremely immature, state. The young find their own way to the mother's ventral pouch, where they attach themselves to the teats until they are sufficiently advanced to emerge and learn to feed themselves.

mastodon An extinct proboscid (q.v.) characterized by having many tooth-cusps rounded like a nipple.

Mesolithic The Middle Stone Age, a stage of human culture occurring, in time, between the Paleolithic (Old Stone Age) and the Neolithic (New Stone Age).

Mesozoic The geological "middle life" era, from about 225 million to about 65 million years ago. Includes the Triassic, Jurassic, and Cretaceous periods.

Miocene The fourth of the five periods of the Tertiary era. Its name ("moderately recent") refers to the proportion of animal species of recent (i.e. modern) type in the fauna of the time.

mollusk A soft-bodied invertebrate, generally with a hard limy protective shell. Most have an extensible muscular foot for moving about; some, such as the mussel, are fixed. Some are very mobile predators, like the octopus. Cuttlefish and slugs have only a vestige of a shell; the octopus has none.

monocotyledon A flowering plant with only a single "seed-leaf", or cotyledon (q.v.). The leaves of monocotyledons tend to be strap-like, with parallel veins, e.g. grasses, irises, orchids, palms (the only woody group).

mosasaur A marine carnivorous reptile of the later Mesozoic, combining the characters of a lizard and of a snake. The name comes from the Latin for the river Meuse, since the fossil remains of the animal were first discovered on that river, at Maastricht in Holland.

natural selection The process whereby those animal individuals and species best fitted for the "struggle for existence" in their particular environment survive the attacks of enemies and rivals, win the fittest mates and ensure the passing on of the special characters that procured their success to their descendants. Those less fitted do not survive to breed, so that disadvantageous variations are not perpetuated in the species.

Neolithic The New Stone Age, preceding the discovery of metals. A level of culture that includes settled habitations, agriculture, stock-breeding, pottery manufacture, and weaving among its important innovations.

notochord The cartilaginous (gristly) rod stiffening the chordate body, replaced in higher chordates (q.v.) by bony segments (vertebrae), which surround the main nerve-chord to form the backbone. Above the notochord is the neurochord, containing the nervous system.

Oligocene The third period in the Tertiary era, in which most of the animal species have since become extinct, but a "few" (Greek *oligoi*) have survived into the Recent (Postglacial) period.

omnivore An animal that can adapt its feeding habits to whatever kind of nourishment, vegetable or animal, is available. Man is the best example.

Ordovician The second period of the Paleozoic era, named after a Welsh tribe of the time of the Roman conquest, in whose former territory the rocks of the period were first recognized.

ostracoderm An early Paleozoic (Ordovician) jawless fish with plates of bony armour covering its skin.

Paleolithic The Old Stone Age, an early stage of human culture marked by the use of flaked stone tools and weapons, whose makers were wandering hunters and gatherers.

paleontology Properly "the study of ancient (living) beings", i. e. including paleobotany with paleozoology.

Paleozoic Era of "ancient life", the first in which fossilized hard parts of contemporary animals are preserved.

pelycosaurs Early (Carboniferous) unspecialized reptiles representing the stem that probably gave rise to the first mammals in the Mesozoic.

period In geology, a subdivision of time secondary to the major eras, e.g. the Carboniferous and Permian are the last two periods of the Paleozoic.

Permian The sixth and final period of the Paleozoic era. Named after the province of Perm, in north European

Russia, where its characteristic red sandstones are very widely displayed.

photosynthesis The process by which green plants, using their contained pigment, chlorophyll, absorb the energy of sunlight to build up glucose (and thence their most complex carbon compounds) from carbon dioxide, water, and mineral salts.

placenta The organ, finally discarded as the afterbirth, which during pregnancy enables the two-way diffusion of gases and nutrients between the bloodstreams of mother and foetus to take place without their mixing directly. All higher mammals have a placenta.

placoderms An extinct class of early (Devonian) vertebrates with primitive jaws, and skins armoured with bony plates.

placodonts Triassic marine reptiles unspecialized save in having flat teeth in the palate, an adaptation to a diet of shellfish.

Pleistocene The next to the last geological period, during which most of the animal species were identical with those still living now, though a few (mostly mammals) have since become extinct. The Pleistocene and Holocene (q.v.) form the Quaternary era, in which we live today.

plesiosaurs Marine reptiles usually with a small head on a long swan-like neck, barrel-shaped body propelled by four oar-like flippers, and a short tail. They are mostly Jurassic.

Pliocene The final period of the Tertiary. The animals of the time included a large number of still surviving species.

Precambrian The era (almost nine times as long as all the rest together) from the earth's origins to the formation of the earliest (Cambrian) fossil-bearing rocks. During the earlier (and greater) part of it, the simplest forms of life must have been developing, but during the later part (Proterozoic), fairly highly organized living things (but without preservable hard parts) must have been present. Most Precambrian "fossils" are still debatable, but the Cambrian shows many highly organized animals that must have had a long earlier history.

primate A member of the order (Primates) of mammals that includes man, as well as apes, monkeys, lemurs, and tree shrews.

proboscids Animals with trunks, members of the order Proboscidea, including mastodons, mammoths, and elephants.

Proterozoic The later part of the Precambrian, during which life must have originated and developed into forms that could leave traces (fossils) in the rocks.

protozoa Simple single-celled animals barely visible with the naked eye. In all likelihood the first animals to be distinguished from plants belonged to this phylum. They must have lived on primitive plants, being unable, like these, to make food by photosynthesis (q.v.).

pterodactyl A flying reptile of the Jurassic and Cretaceous in which the wing-membrane was supported in part by the relatively short arms and legs but for the greater part by the enormously elongated fourth finger, which formed its leading edge.

pterosaurs An extinct order of flying (or more likely gliding) reptiles of the later Mesozoic.

Quaternary The last geological era, comprising the Pleistocene and Holocene periods. In time probably corresponds to the last two to three million years.

reptile Animals that are cold-blooded, dry-skinned, and lay eggs with membranous or horny shells; they are active only in warm climates and seasons.

saurian A general term, without exact taxonomic status, for any lizard-like reptile.

seed The embryo of a plant, generally together with a varying amount of nutrient stored for it by its parents, enclosed in a tough seed-coat, skin or shell, enabling it to lie dormant, sometimes for years, through dry, cold, or dark conditions.

Silurian The third period of the Paleozoic era. First named in Wales after an ancient British tribe called the Silures by the Romans.

species A group of animals that can interbreed to give fertile offspring. There are exceptions. One or more species that show similarities are grouped together as a genus.

spore A single tiny cell released into air or water (together with thousands like it) by a lower plant; if it lands in a suitable environment, it germinates to produce a plantlet called prothallus, the sexual stage of its reproduction.

Separate male and female cells produced by prothalli meet to form an embryo from which will grow another spore-bearing individual.

stegocephalian One of a group of Devonian-Triassic amphibians, presumed ancestors of the reptiles.

strata (singular: **stratum**) The successive layers in which sedimentary rocks are laid down, the younger covering their older predecessors, through the agencies of moving ice, water, or wind.

subspecies Generally a population of a species which, geographically isolated from the main body of its peers, has lost reproductive contact with them, and, by inbreeding, has in time developed minor variations in colouring, plumage, or other characters that distinguish it from typical members of its species.

taxonomy The theoretical study of classification, including its bases, principles, procedures, and rules.

Tertiary The third of the main geological time eras, comprising the Paleocene, Eocene, Oligocene, Miocene, and Pliocene periods.

testudinate A member of the land-living branch of the reptilian order Chelonia, which comprises both tortoises and turtles.

thecodonts Members of an extinct Triassic order of small, carnivorous, often biped, reptiles characterized by having rooted teeth in sockets in the jaw.

Triassic The first period of the Mesozoic, so called because, in Europe, it is distinctively divided into three series of rock strata.

trilobite An early Paleozoic marine arthropod (q.v.), its body characteristically divided into three "lobes" by two longitudinal grooves crossing the many segments.

ungulate A mammal with hooves, such as horses, cattle, camels.

vascular system The system of vessels that conveys liquids and solutions from one part of a plant or animal body to another.

ventral The belly- or under-side of an animal or animal-structure. In a standing man the ventral side is in front. The term is thus independent of attitude.

vertebrate An animal with a backbone. A member of a main subdivision of the phylum Chordata.

Index

Page numbers in *italics* refer to illustrations, diagrams, or their captions